W. Dietmaier   C. Wittwer   N. Sivasubramanian (Eds.)

Rapid Cycle Real-Time PCR – Methods and Applications
Genetics and Oncology

Springer

*Berlin
Heidelberg
New York
Barcelona
Hong Kong
London
Milan
Paris
Tokyo*

Table of Contents

**Introduction for Genetics and Oncology Volume
Rapid Cycle Real-Time PCR**
Methods and Applications 1
NATARAJAN SIVASUBRAMANIAN

Part I
Methods Useful in Genetics and Oncology

**Quantification of Cytokine mRNAs in Human Myocardial Biopsy Samples
by Real-Time Quantitative PCR Technology Using the LightCycler Instrument** ... 5
XI ZHU, GEORG BAUMGARTEN, FENG WANG, ZIAD DIBBS,
ABHINAV DIWAN, GUILLERMO TORRE-AMIONE, DOUGLAS L. MANN,
NATARAJAN SIVASUBRAMANIAN

**Quantitative Two-Step RT-PCR for the Detection of Human ABCA1 Transporter
on LightCycler Using Hybridization Probes and External Standards** 15
DANUTA KIELAR, WOLFGANG DIETMAIER, THOMAS LANGMANN,
CHARALAMPOS ASLANIDIS, MARIO PROBST, MAREK NARUSZEWICZ,
GERD SCHMITZ

Quantification of Human Genomic DNA Using Retinoic X Receptor B Gene 27
NATHALIE PIERI-BALANDRAUD, JEAN ROUDIER, CHANTAL ROUDIER

**Genotyping by Guanosine-Dependent Quenching
of Single-Labeled Fluorescein Probes** 35
ANDREW O. CROCKETT

**Limitations of Melting Curve Analysis Using SYBR Green I –
Fragment Differentiation and Mutation Detection in the CFTR-Gene** 47
S. KLEINLE, K. TABITI AND S. GALLATI

**SYBR Green I Analysis of the Trinucleotide Repeat Responsible
for Huntington's Disease** 57
CAMERON N. GUNDRY, CARL T. WITTWER

Part II
Applications in Genetics

Parallel Genotyping of Different Genes: A Rapid Real-Time PCR Approach 67
STEFAN FRONHOFFS, THOMAS BRÜNING, HANS VETTER, YON KO

Detection of a Single Base Substitution in Single Cells by Melting Peak Analysis Using Dual-Color Hybridization Probes 77
GERARD PALS

Rapid Screening for Five Major Cystic Fibrosis Mutations by Melting Peak Analysis Using Fluorogenic Hybridization Probes 85
SIEGFRIED BURGGRAF, NAEEM MALIK, EDITH SCHUHMACHER, BERNHARD OLGEMÖLLER

LightCycler PCR for the Polymorphisms −308 and −238 in the *TNF Alpha* Gene and for the TNFB1/B2 Polymorphism in the *LT Alpha* Gene 95
LUKAS BESTMANN, NÄDER HELMY, FELICIA GAROFALO, AYNUR DEMIRTAS, DIETER VONDERSCHMITT, FRIEDRICH E. MALY

Rapid Genotyping of 2-bp and 9-bp Deletion Mutations Using the LightCycler 107
TSUTOMU AOSHIMA, MITSUHARU KAJITA, YOSHITAKA SEKIDO, SHUNJI MIMURA, KAZUYOSHI WATANABE, KAORU SHIMOKATA, TOSHIMITSU NIWA

Genotyping of the Methionine-Valine Polymorphism at Codon 129 of the Human Prion Protein by Melting Point Analysis of Fluorescently Labeled Hybridization Probes 115
SIEGFRIED BURGGRAF, SIEGFRIED KÖSEL, SABINE LOHMANN, REINHARD BECK, BERNHARD OLGEMÖLLER

Rapid Detection of Missense Mutations in the Prostatic Steroid 5α-Reductase Gene Using Real-Time Fluorescence PCR and Melting Curve Analysis 129
MARKUS NAUCK, WINFRIED MÄRZ, HEINRICH WIELAND

Part III
Applications in Oncology

Analysis of Microsatellite Instability by Melting Peak Analysis with BAT26 and BAT25 Specific Fluorescence Hybridization Probes 139
WOLFGANG DIETMAIER, ARNDT HARTMANN, FERDINAND HOFSTÄDTER

W. Dietmaier   C. Wittwer   N. Sivasubramanian (Eds.)

Rapid Cycle Real-Time PCR – Methods and Applications

Genetics and Oncology

With 70 Figures and 107 Tables

Springer

Dr. Wolfgang Dietmaier
Institut für Pathologie
Franz-Josef-Strauß-Allee 11
93053 Regensburg
Germany

Professor Dr. Carl Wittwer
Department of Pathology
University of Utah Medical School
Salt Lake City, UT 84132
USA

Dr. Natarajan Sivasubramanian
Baylor College of Medicine
2002 Holcombe Blvd.
VAMC Building 110, Room 165
Houston, TX 77030
USA

ISBN 3-540-42600-0   Springer Verlag Berlin Heidelberg New York

Library of Congress applied for
Rapid cycle real-time PCR : methods and applications ; genetics and oncology / Wolfgang Dietmaier, Carl Wittwer, Natarajan Sivasubramanian (eds.). p. cm.
    Includes bibliographical references.
    ISBN 3540426000 (alk. paper)
1. Polymerase chain reaction. 2. Molecular genetics–Methology. 3. Cancer–Genetic aspects. 4. Molecular probes. I. Dietmaier, Wolfgang, 1963 – II. Wittwer, C. (Carl), 1955 – III. Sivasubramanian, Natarajan.

This work is subject to copyright. All rights are reserved, whether the whole or part of the material is concerned, specifically the rights of translation, reprinting, reuse of illustrations, recitation, broadcasting, reproduction on microfilm or in any other way, and storage in data banks. Duplication of this publication or parts thereof is permitted only under the provisions of the German copyright Law of September 9, 1965, in its current version, and permission for use must always be obtained from Springer-Verlag. Violations are liable for prosecution under the German Copyright Law.

Springer-Verlag Berlin Heidelberg New York
a member of BertelsmannSpringer Science+Business Media GmbH

http://www.springer.de/medizin

© Springer-Verlag Berlin Heidelberg 2002
Printed in Germany

The use of general descriptive names, registered names, trademarks, etc. in this publication does not imply, even in the absence of a specific statement, that such names are exempt from the relevant protective laws and regulations and therefore free for general use.

Product liability: The publisher cannot guarantee the accuracy of any information about dosage and application thereof contained in this book. In every individual case the user must check such information by consulting the relevant literature.

Cover Design: design & production, 69121 Heidelberg, Germany
Production: ProEdit GmbH, 69126 Heidelberg, Germany
Typesetting: TBS, 69207 Sandhausen, Germany
Printed on acid free paper   SPIN 10839003   18/3130 Re – 5 4 3 2 1 0

Two Color Multiplexing and Typing of Human Papillomavirus Types 16, 18 and 45 on LightCycler 147
Monica L. Henriquez, Brian E. Caplin, Randy P. Rasmussen

Quantitative Analysis of AML1-ETO Fusion Transcripts in t(8;21) Positive AML Using Real-Time RT-PCR 159
Martin Weisser, Claudia Schoch, Torsten Haferlach, Wolfgang Hiddemann, Susanne Schnittger

Rapid Quantitative Detection of Free Cancer Cells in the Peritoneal Cavity of Gastric Cancer Patients with Real-Time CEA RT-PCR Using Hybridization Probes 169
Hayao Nakanishi, Yasuhiro Kodera, Masae Tatematsu

Quantitative Measurement of the mRNA Expression of the Tumor-Associated Antigen PRAME by Real-Time RT-PCR Using LightCycler and SYBR Green I Technology 177
Jochen Greiner, Mark Ringhoffer, Anita Szmaragowska, Sandra Hübsch, Hartmut Döhner, Michael Schmitt

Expression Analysis of Telomerase-Genes hTERT and hTR by Quantitative PCR on LightCycler 187
Bernd Frodermann, Christopher Poremba

Measurement of *MDR1* Gene Expression by Real-Time Quantitative RT-PCR Using the LightCycler Instrument 199
Chung-Che Chang, Sherrie Perkins, Carl Wittwer

Abbreviations

Bp	Base pair
EDTA	Ethylenediamine Tetraacetic Acid
F	Fluorescein
LCRed640	LightCycler Red 640 Fluorescent Dye
LCRed705	LightCycler Red 705 Fluorescent Dye
Nt	Nucleotide(s)
P	Phosphate
PBS	Phosphate Buffered Saline
SDS	Sodium Dodecylsulfate
TE	1xTE Buffer Contains 0.01 M Tris, 0.001 M EDTA, pH 8.0
TE'	1xTE' Buffer Contains 0.01 M Tris, 0.0001 M EDTA, pH 8.0
TEN (=STE)	1xTEN Buffer Contains 0.01 M Tris (pH 8.0), 150 mM NaCl, 0.05% Tween 20
$T_m(°C)$	Melting Temperature
n-n	nearest neighbor
FRET	Fluorescence Resonance Energy Transfer
ΔH	Enthalpy
ΔS	Entropy
ΔG	Free Energy
TPMT	Thiopurine Methyltransferase
XO	Xanthine Oxidase
AA	Amino Acid
ORF	Open Reading Frame

LightCycler is a trademark of a member of the Roche Group
MagNA Pure LC is a trademark of a member of the Roche Group

Introduction for Genetics and Oncology Volume Rapid Cycle Real-Time PCR
Methods and Applications

The variant expression of genes and their protein products is one of the fundamental mechanisms of pathogenesis in human disease. Variant expression of genes includes both altered levels of normal gene products and modified proteins resulting from mutations in exons. In either case, detection of variant expression during early pathogenesis can provide a rational therapeutic approach.

Recently, a new technology has been developed to detect variant gene expression based on the polymerase chain reaction. Desired segments of nucleic acids are amplified and quantified very quickly in real-time with great accuracy. This technology combines rapid temperature cycling with fluorescent detection of products and has become known as, "Rapid Cycle Real-Time PCR". The methods described in this volume address detection of mutations and mRNA quantification of a variety of target genes. In addition, to assess the stability of the genome by analysis of microsatellite sequences are presented. Mutation detection methods usually use hybridization probes, whereas for mRNA quantification, both SYBR Green I and hybridization probes are commonly used.

The LightCycler (Roche Diagnostics) used by the contributors to this volume was the first system that permitted the investigator to evaluate target gene amplification after each cycle in real time. Recently, an automated nucleic acid extraction system, the MagNAPure LC (Roche Diagnostics), was introduced to complement the LightCycler system. The combination of MagNAPure LC and LightCycler now permits the complete automation of extraction, purification, concentration, amplification and detection of target nucleic acid from patient samples. In addition, immediate melting curve analysis after PCR with hybridization probes or PCR products is extraordinarily useful. In this volume, the editors Dr. Dietmaier, Professor Wittwer and Dr. Sivasubramanian, have used this feature to determine the stability of the genome in microsatellite regions. This innovative use aids in the diagnosis of hereditary non-polyposis colorectal cancers.

Real time PCR technology also has many applications in human heart failure research, diagnosis and therapy. For example, current therapeutic treatment of human heart failure is based mostly on physiological measurements of heart function. However, physicians and researchers are handicapped because of the inability to determine the biochemistry of the heart during therapeutic treatment. As a first step to meet this deficiency, Zhu et al. in this volume present a method to quantify mRNAs in myocardial biopsy samples of 1 mg size. This method was recently used to understand the biochemical basis of hebernating myocardium in patients (1). Thus, future development of this technology should eventually help in the diagnostics and treatment of human heart failure.

Rapid cycle real time PCR is a revolutionary testing platform for the contemporary genetics laboratory. This technology is changing the way we identify and quantify variant gene expression. I am certain you will find the information provided herein very useful.

April 2002 NATARAJAN SIVASUBRAMANIAN for the Editors

1) Kalra DK, Zhu X, Ramchandani MK, Lawrie G, Reardon MJ, Lee-Jackson D, Winters WL, Sivasubramanian N, Mann DL, Zoghbi WA. (2002). Increased myocardial gene expression of tumor necrosis factor-alpha and nitric oxide synthase-2: a potential mechanism for depressed myocardial function in hibernating myocardium in humans. Circulation;105(13):1537–40

Methods Useful in Genetics and Oncology I

Quantification of Cytokine mRNAs in Human Myocardial Biopsy Samples by Real-Time Quantitative PCR Technology Using the LightCycler Instrument ... 5
Xi Zhu, Georg Baumgarten, Feng Wang, Ziad Dibbs, Abhinav Diwan, Guillermo Torre-Amione, Douglas L. Mann, Natarajan Sivasubramanian

Quantitative Two-Step RT-PCR for the Detection of Human ABCA1 Transporter on LightCycler Using Hybridization Probes and External Standards 15
Danuta Kielar, Wolfgang Dietmaier, Thomas Langmann, Charalampos Aslanidis, Mario Probst, Marek Naruszewicz, Gerd Schmitz

Quantification of Human Genomic DNA Using Retinoic X Receptor B Gene 27
Nathalie Pieri-Balandraud, Jean Roudier, Chantal Roudier

Genotyping by Guanosine-Dependent Quenching of Single-Labeled Fluorescein Probes 35
Andrew O. Crockett

Limitations of Melting Curve Analysis Using SYBR Green I – Fragment Differentiation and Mutation Detection in the CFTR-Gene 47
S. Kleinle, K. Tabiti and S. Gallati

SYBR Green I Analysis of the Trinucleotide Repeat Responsible for Huntington's Disease 57
Cameron N. Gundry, Carl T. Wittwer

Quantification of Cytokine mRNAs in Human Myocardial Biopsy Samples by Real-Time Quantitative PCR Technology Using the LightCycler Instrument

Xi Zhu, Georg Baumgarten, Feng Wang, Ziad Dibbs, Abhinav Diwan, Guillermo Torre-Amione, Douglas L. Mann, Natarajan Sivasubramanian*

Introduction

Understanding the pathophysiology of human heart failure [1–3] necessitates the sensitive measurement of gene expression in human myocardium. Estimating the quantities of mRNAs in an explanted failing human heart sample after heart transplantation is a relatively easy task because of the availability of large-sized tissue samples. However, when the availability of tissue samples is very low, especially from patients who are undergoing therapeutic treatment [4], estimating the quantities of mRNAs requires a highly sensitive method of measurement. In this study, we have developed a highly sensitive method for the quantification of cytokine mRNAs in human myocardial biopsy tissue samples as low as 2–3 mg. This method involves isolation of total RNA from frozen biopsy tissue samples, capturing and immobilizing poly-A RNA, reverse transcription of RNA to obtain single-stranded complementary DNA, and quantification of cytokine mRNAs using a real-time quantitative polymerase chain reaction method.

Materials

Equipment

Biometra Gradient Thermocycler (Göttingen, Germany)
LightCycler and data analysis software version 3.0 (Roche Diagnostics, Indianapolis, IN, USA)
Beckman 540 Spectrophotometer
Dataminder 3.2 software for primer design (Karen Usdin, NIDDK, NIH, MA, USA)
Polytron 3000 (Brinkmann Instruments, Inc., Westbury, NY, USA)

Reagents

Oligonucleotides (InVitrogen, San Diego, CA, USA)
Qiagen plasmid purification kits (Qiagen, Germany)
RNA STAT-60 (Tel-Test Inc., USA)
TRI$_{ZOL}$ Reagent (InVitrogen)

---

* N. Sivasubramanian (✉) (e-mail: nats@bcm.tmc.edu)
  Winters Center for Heart Failure Research, Cardiology Section of the Department of Medicine, Veterans Administration Medical Center, Baylor College of Medicine, Houston, TX 77030, USA

mRNA Capture Kit (Cat. 1787896; Roche Molecular Biochemicals, USA)
1st Strand cDNA Synthesis Kit for RT-PCR (Cat. 1483188; Roche Molecular Biochemicals)
LightCycler DNA Master SYBR Green I Kit (Cat. 2015099; Roche Molecular Biochemicals)
Betaine 5 M (Cat. B0300; Sigma, USA)
TNF-α cDNA clone (American Type Culture Collection Cat. 39918)
IL-1β cDNA clone (American Type Culture Collection Cat. 581769)
IL-6 cDNA clone (American Type Culture Collection Cat. 567519)
β-actin cDNA (Cat. 9800-1, Clontech Laboratories, USA)

Procedure

Sample Preparation

Human myocardial biopsy tissue samples (~2–3 mg and >50 mg) from failing hearts were collected postmortem after heart transplantation, flash frozen in liquid nitrogen, and stored at –80°C. The frozen samples were thawed in 1 ml of Trizol or RNA-STAT solution and immediately homogenized using a Polytron 3000 homogenizer at 30,000 rpm for 30 s. After homogenization, the total RNA was extracted and precipitated according to the manufacturer's protocol. Isolated RNA was immediately stored at –80°C until use. Due to the minute amount of total RNA from myocardial biopsy samples (~2–3 mg), the concentration of RNA was not spectrophotometrically determined. However, when the tissue samples were of a larger size (>50 mg), the concentration and purity of RNA preparations were measured at A_{260}, A_{280} and A_{320} using a Beckman 540 spectrophotometer.

Purification of Poly-(A) RNA and cDNA Synthesis

Poly-A RNA from the total RNA was captured using biotin-labeled oligo-(dT)$_{20}$ (mRNA Capture Kit) according to the manufactures protocol. Following capture, the poly-A RNA was immobilized by transferring the captured RNA to straptavidin-coated microfuge tubes. After immobilization, the solution was removed from the straptavidin-coated tubes and the tubes were washed three times and stored at –80°C. First-strand cDNA synthesis (1st Strand cDNA Synthesis Kit for RT-PCR) was directly performed on the immobilized poly-A RNA according to the manufacturer's protocol. Briefly, oligo-p(dN)$_6$ primer was added to the straptavidin-coated tubes containing immobilized poly-A RNA and reverse transcribed into single-stranded cDNA. Because of the use random hexamer for first strand cDNA synthesis, multiple copies of the same mRNA will be made as short single-stranded cDNA segments from different locations, and these nascent cDNA segments can be freed from the template by denaturation. Thus, the reaction was terminated by denaturing the cDNA products at 95°C, and the synthesized cDNA products were removed from the straptavidin-coated tubes and stored at –20°C for quantitative PCR. Later, the straptavidin-coated tubes containing the immobilized poly-A RNA were washed again to remove the residual cDNA and stored at –80°C for future use. We have used the immobilized poly-A RNA three times for repeated synthesis of cDNA, and quantitative PCR yielded identical results each time. Additionally, we have also generated single-stranded cDNA products directly from the total RNA using a random hexamer method without isolating poly-A RNA.

Reverse transcription reaction for each 20-µl reaction:

	Volume [µl]	[Final]
25 mM MgCl$_2$	4	5 mM
Deoxynucleotide	2	1 mM
Random primer p(dN)$_6$	2	0.08 A$_{260}$ U
RNase inhibitor	1	50 U
AMV reverse transcriptase	0.8	20 U
10× Reaction Buffer	2	1×
Milli-Q H$_2$O	8.2	
Total master mix volume per reaction	20	
RT reaction	25°C	10 min
	42°C	60 min
	99°C	5 min

The LightCycler protocol was as follows:
- Denaturation for 30 s at 95°C
- Amplification

Parameter	Value			
Cycles	55			
Type	Quantification			
	Segment 1	Segment 2	Segment 3	Segment 4
Target temperature [°C]	95	55–60	72	81–83
Incubation time [s]	0	5	10	5
Temperature transition rate [°C/s]	20	20	20	20
Acquisition	None	None	None	Single

- Melting Curve Analysis

Parameter	Value		
Cycles	1		
Type	Melting curve analysis		
	Segment 1	Segment 2	Segment 3
Target temperature [°C]	95	60	95
Incubation time [s]	0	30	0
Temperature transition rate [°C/s]	20	20	0.1
Acquisition	None	None	Cont.

The fluorimeter gain of Channel 1 was set to 1.

PCR Primer Design

Primers for quantitative PCR were designed to satisfy the following criteria: Firstly, primers were designed so that there is no complementarity between the sense and antisense PCR primers, especially at the 3′ end, thus avoiding the formation of primer-dimers during amplification. This is particularly important since primer-dimers are formed in samples containing low amounts of target gene mRNAs. Moreover, the abundant formation of primer-dimers in the early cycles of PCR limits the availability of primers for amplification in the late cycles. Secondly, the melting temperature (T_m) of the gene-specific product as well as primer-dimers was taken into consideration. The PCR primers were designed so that the gene-specific PCR product T_m exceeds the primer-dimer T_m by at least 10°C. The GC content of the gene-specific PCR product and primers was maintained around 50%. Additionally, PCR sense and antisense primers were designed in such a way that they are positioned either on the intron–exon junction or on the different exons of the respective target genes. With these criteria, the PCR primers for TNF-α, IL-1β, and IL-6 (Table 1), as well as for the housekeeping gene β-actin, were designed using the DataMinder 3.2 computer program (Mac software kindly provided by Dr. Karen Usdin at NIH). The T_m of the primers was between 49–68°C. The T_m of the amplicons was 85°–88°C. The amplicon lengths were 180–300 bp.

Generation of Copy Number cDNA Standards for Quantitative PCR

For generating copy number cDNA standards, cDNA clones encoding the target genes for TNF-α, IL-1β, IL-6, and β-actin were obtained. Plasmids containing cDNA inserts were purified from 50-ml bacterial cultures according to the manufacturer's protocol (Qiagen, Germany). Following the confirmation of the cDNA

Table 1. Oligonucleotides

Human Cytokine Coding Genes [(GenBank Accession Numbers, TNFα (M10988); IL-1β (M15330); IL-6 (M29150); β-Actin (X00351)]			
	Length	GC (%)	T_m (°C)
TNF-α			
Primers			
5'-AAGAGTCCCCCAGGGACCTCT-3'	21	61	53
5'-CCTGGGAGTAGATGAGGTACA-3'	21	52	49
Product	230		88.2
IL-1β			
Primers			
5'-GCCCTAAACAGATGAAGTGCTC-3'	22	50	60.8
5'-AGAAGGTGCTCAGGTCATTCTC-3'	22	50	60.8
Product	198		86.2
IL-6			
Primers			
5'-CAGCTATGAACTCCTTCTCCACAAGC-3'	26	50	64.8
5'-CTGAGATGCCGTCGAGGATGTACCG-3'	25	60	67.9
Product	201		85.5
β-Actin			
Primers			
5'-AGCACGGCATCGTCACCAACT-3'	21	57.1	62.7
5'-TGGCTGGGGTGTTGAAGGTCT-3'	21	57.1	62.7
Product	180		86.0

sequence by automated sequencing, the plasmids were digested with the appropriate restriction enzymes to generate the copy-number template standards for cytokine mRNAs. The expected fragment sizes were confirmed by agarose gel-electrophoresis and the DNA fragments were eluted according to the manufacturer's protocol (Qiagen). The concentrations of the cDNA fragments were determined spectrophotometrically. Copy numbers of cDNAs were empirically determined using the following equation: copy number=(grams/MW) × Avogadro constant (Avogadro constant=6.0221367×10^{23}). The quantified cDNA fragments were later serially diluted in log increments and were used as known copy-number template standards in quantitative PCR.

Results

Before conducting the quantitative PCR, the synthesized PCR primers were first tested in a conventional gradient thermocycler (Biometra) to determine the optimal cycling conditions for generating appropriate gene-specific PCR products. This is especially important since PCR amplification can occur at a wide variety of conditions. The optimal annealing temperature for primers was chosen based on the maximal yield of gene-specific PCR product with very low or no formation of primer-dimers or nonspecific products. Later, these optimal PCR conditions were imported into the LightCycler for conducting quantitative PCR. The first 5–6 capillaries contained an empirically determined appropriate known copy number standard cDNA serially diluted (1–10^{-6}) with water from a stock solution. A negative control with no template cDNA was also included. The range for the copy number cDNA standards for the respective cytokines was chosen based on its expected amount in the myocardium. Each unknown cDNA sample was measured twice with two dilutions. The PCR reaction mixtures were assembled first and later transferred into precooled capillaries, placed into special adapters, and quickly centrifuged to bring the reaction mixture to the bottom of the capillaries. After placing the capillaries into the rotor of the LightCycler instrument, quantitative PCR cycling was initiated and the real-time amplification was monitored (panel A in Figs. 1–4) using the LightCycler software.

Quantification of Cytokine mRNA Using Real-Time Quantitative PCR

The SYBR Green Master Mix for each 20-μl reaction was:

	Volume [μl]	[Final]
LC-DNA Master SYBR Green I	2	1×
MgCl (25 mM)	2.4	5 mM
Primers (10 μM each)	1+1	0.5 μM each
Taq antibody	0.6	0.21 μM
Betaine 5 M	4.8	1.2 M
Total master mix volume per reaction	15–19	

In total, 15–19 μl of master mix and 1–5 μl cDNA were added to each capillary. Sealed capillaries were centrifuged and place into the LightCycler rotor.

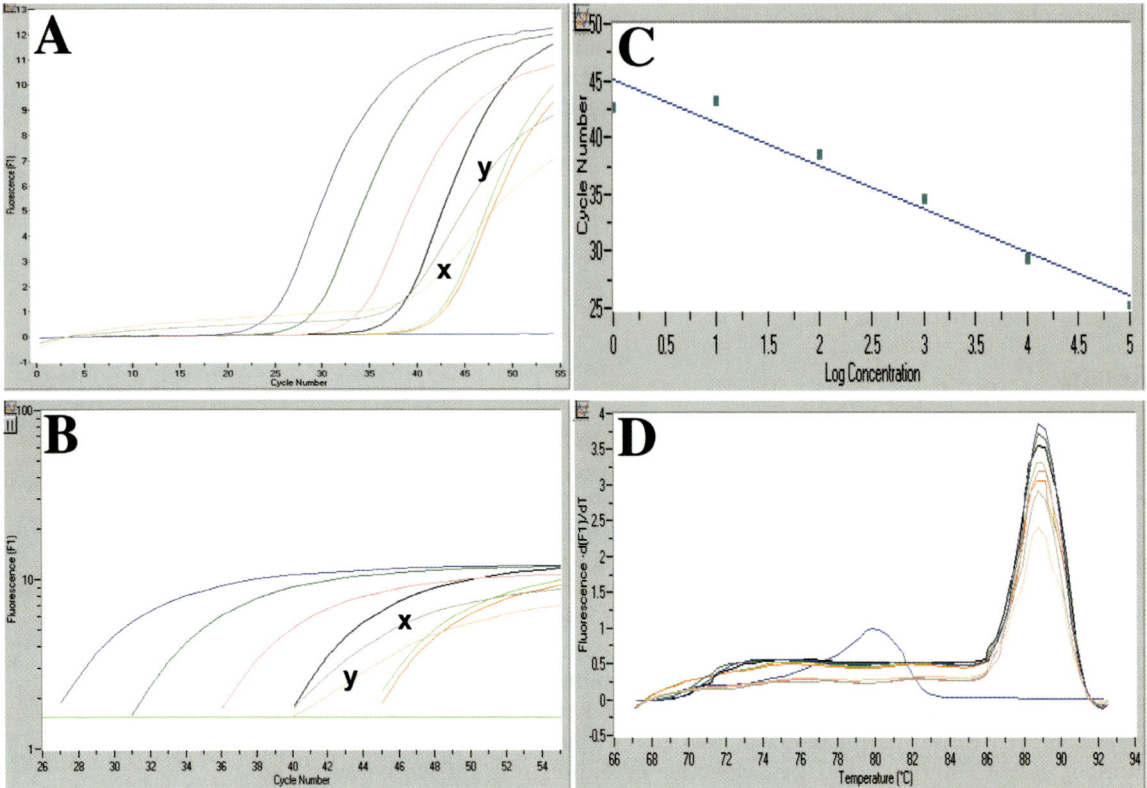

Fig. 1A–D. Quantitative PCR of TNF-α. **A** Monitoring of entire SYBR Green I PCR reaction. Fluorescence vs. cycle number plot of TNF-α cDNA amplification of $10–10^5$ copies of known standard and two unknown samples (*black arrows*). **B** Identification of log-linear cycles for quantification; *green line* denotes crossing line. Log fluorescence vs. cycle number plot of TNF-α cDNA amplification. No-template negative control (*blue line*) and $10–10^5$ copies of TNFα cDNA standard (1 copy=*green*; 10 copies=*red*; 10^2 copies=*black*; 10^3 copies=*pink*; 10^4 copies=*dark green*; 10^5 copies=*dark blue*) and two unknown myocardial biopsy samples (*x* and *y*). **C** Standard curve of known copy number standards for TNF-α (slope: –4.824; intercept: 59.35; error: 0.258; *r*: –0.99). **D** Melting curve analysis of PCR products from all reactions T_m (80.0°C) of non-specific primer-dimers product is lower than T_m (88.2°C) of gene-specific products

After the PCR amplification, melting curve analysis (panel D in Figs. 1–4) was done first to validate the generation of the expected gene-specific PCR product. Measuring the T_m of various PCR products distinguished the gene specific PCR product from artifact "primer-dimers" or any other nonspecific PCR product. Since the T_m of the gene-specific PCR product differs from primer-dimers by at least 10°C because of our primer design, the peak of the gene-specific product did not overlap with the primer-dimer peak and thus the accuracy of the quantification was validated. Log-linear PCR cycles (green crossing line in panel B in Figs. 1–4) were identified collectively for all the tested samples and the fit point analysis method was

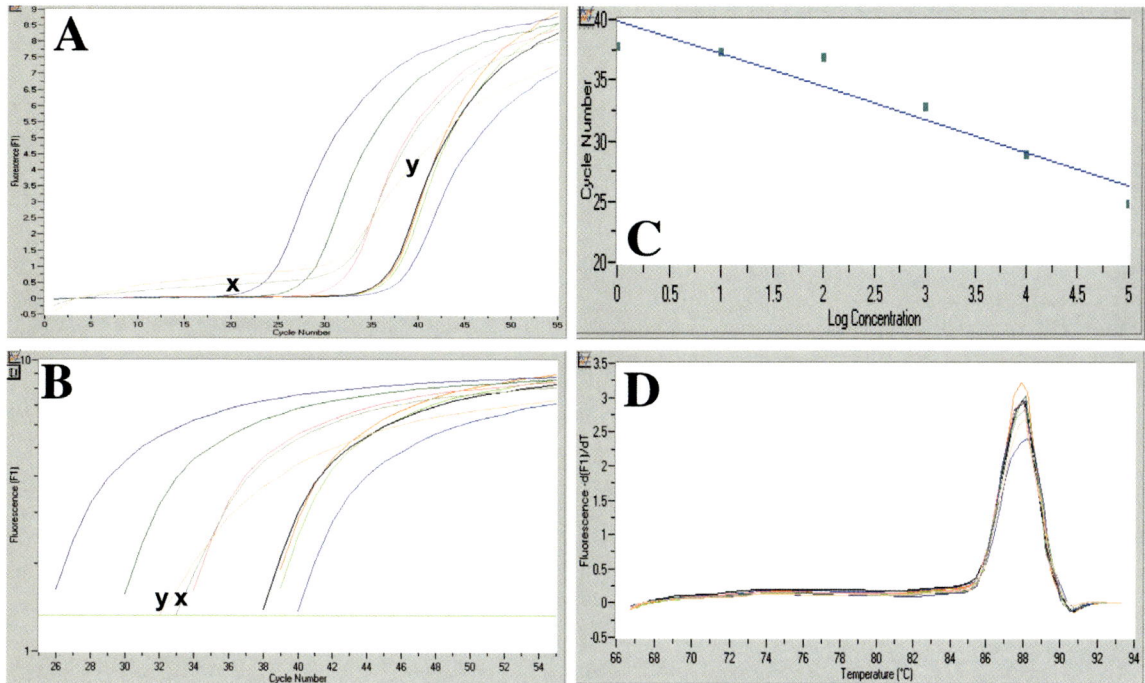

Fig. 2A–D. Quantitative PCR of IL-1β. **A** Monitoring of entire SYBR Green I PCR reaction. Fluorescence vs. cycle number plot of IL-1β cDNA amplification of $10-10^5$ copies of standard and two unknown samples (*black arrows*). **B** Identification of log-linear cycles for quantification; *green line* denotes crossing line. Log fluorescence vs. cycle number plot of IL-1β cDNA amplification. No-template negative control and $10-10^5$ copies of IL-1β cDNA standard (1 copy=*green*; 10 copies=*red*; 10^2 copies=*black*; 10^3 copies=*pink*; 10^4 copies=*dark green*; 10^5 copies=*dark blue*) and two unknown myocardial biopsy samples (*x* and *y*). **C** Standard curve of known copy number standards for IL-1β (slope: –2.833; intercept: 37.45; error: 0.527; *r*: –0.98). **D** Melting curve analysis of PCR products from all reactions

used for quantification. A standard curve with the known copy number samples was generated (panel C in Figs. 1–4) by adjusting the green crossing line manually so that the error value was very low and the *r* value (correlation coefficient) was close to 1. At this point, the final quantification of the unknown samples was determined. This approach was used to determine the quantity of cytokine (TNF-α, IL-1β, IL-6) and β-actin mRNAs in unknown experimental samples (x and y). Because of the smaller size of the unknown samples, the quantification of cytokine mRNAs was normalized to β-actin mRNA copies. The results are shown in Table 2. The values obtained by this method were also compared with larger size samples (also from X and Y) where the concentration of total RNA was known. These results (data not shown) were exactly identical to the results obtained from small-sized myocardial biopsy samples (Table 2). Thus, using small quantities of myocardial biopsy samples (~2–3 mg), cytokine mRNAs can be quantified.

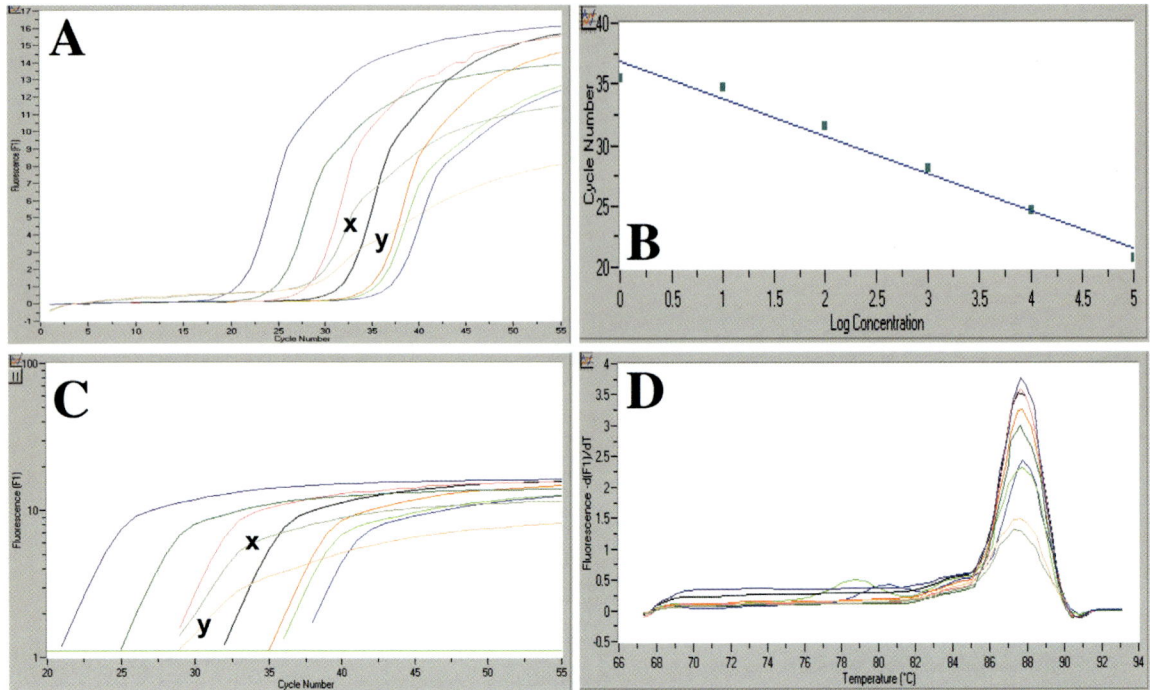

Fig. 3A–D. Quantitative PCR of IL-6. **A** Monitoring of entire SYBR Green I PCR reaction. Fluorescence vs. cycle number plot of IL-1β cDNA amplification of $10-10^5$ copies of standard and two unknown samples (*black arrows*). **B** Identification of log-linear cycles for quantification; *green line* denotes crossing line. Log fluorescence vs. cycle number plot of IL-6 cDNA amplification. No-template negative control (*blue line*) and $10-10^5$ copies of IL-6 cDNA standard (1 copy=*green*; 10 copies=*red*; 10^2 copies=*black*; 10^3 copies=*pink*; 10^4 copies=*dark green*; 10^5 copies=*dark blue*) and two unknown myocardial biopsy samples (*x* and *y*). **C** Standard curve of known copy number standards for IL-6 (slope: −3.1; intercept: 33.94; error: 0.233; *r*: −0.99). **D** Melting curve analysis of PCR products from all reactions

Table 2. Quantification of cytokine mRNAs in myocardial biopsy samples

	Sample X		Sample Y	
Copy number	Absolute	Relative	Absolute	Relative
		/10^6 β-actin mRNA		/10^6 β-actin mRNA
TNF-α	81	24	29	5
IL-1β	267	60	238	42
IL-6	688	200	297	52

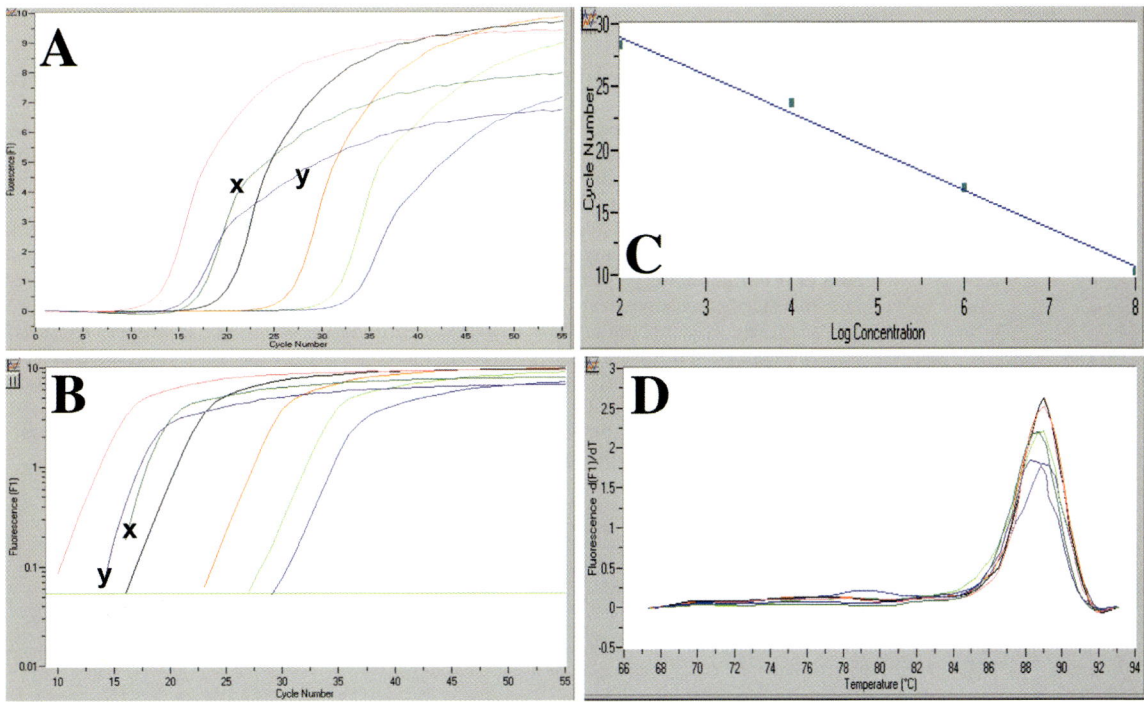

Fig. 4A–D. Quantitative PCR of β-actin. **A** Monitoring of entire SYBR Green I PCR reaction. Fluorescence vs. cycle number plot of IL-1β cDNA amplification of 10^2–10^8 copies of standard and two unknown samples (*black arrows*). **B** Identification of log-linear cycles for quantification – *green line* denotes crossing line. Log fluorescence vs. cycle number plot of β-actin. No-template negative control and 10^2–10^8 copies of β-actin cDNA standard (10^2 copy=*green*; 10^4 copies=*red*; 10^6 copies=*black*; 10^8 copies=*pink*) and two unknown myocardial biopsy samples (x=5.7×10^6; y=3.4×10^6 copies). **C** Standard curve of known copy number standards for β-actin (slope: –3.207; intercept: 36.25; error: 0.228; r: –1.00). **D** Melting curve analysis of PCR products from all reactions

Comments

The proposed method to quantify cytokine mRNAs in a myocardial biopsy sample as low as 2 mg can be applied to a wide range of genes for various cardiovascular diseases [4] (H. Oral et al., in preparation). Using this method, cytokine and iNOS mRNAs were recently quantified in human biopsy samples (2–3 mg) from hibernating myocardium [4]. Additionally, with the use of the mRNA capture method mentioned here, poly-A RNA could be immobilized from these minute tissue samples and could be repeatedly used for mRNA quantification studies. One of the advantages of this approach is that one can accumulate a large amount of knowledge on the myocardial gene expression using a small amount of tissue sample. It does not require a new patient biopsy sample each time. Therefore, using this method the pathophysiology, diagnosis, and treatment of heart diseases can be studied at the gene expression level [5].

References

1. Mann DL (2001) Recent insights into the role of tumor necrosis factor in the failing heart. Heart Fail Rev 6:71–80
2. Nagueh SF, Stetson SJ, Lakkis NM, Killip D, Perez-Verdia A, Entman ML, Spencer WH 3rd, Torre-Amione G (2001) Decreased expression of tumor necrosis factor-alpha and regression of hypertrophy after nonsurgical septal reduction therapy for patients with hypertrophic obstructive cardiomyopathy. Circulation 103:1844–1850
3. Kubota T, Miyagishima M, Alvarez RJ, Kormos R, Rosenblum WD, Demetris AJ, Semigran MJ, Dec GW, Holubkov R, McTiernan CF, Mann DL, Feldman AM, McNamara DM (2000) Expression of proinflammatory cytokines in the failing human heart: comparison of recent-onset and end-stage congestive heart failure. J Heart Lung Transplant 19:819–824
4. Kalra DK, Zhu X, Ramchandani MK, Lawrie G, Reardon MJ, Lee-Jackson D, Winters WL, Sivasubramanian N, Mann DL, Zoghbi WA (2002) Increased myocardial gene expression of TNF and NOS2: a potential mechanism for depressed myocardial function in hibernation myocardium in man. Circulation 105:1537–1540
5. Baumgarten G, Knuefermann P, Mann DL (2000) Cytokines as emerging targets in the treatment of heart failure. Trends Cardiovasc Med 10:216–223

Quantitative Two-Step RT-PCR for the Detection of Human ABCA1 Transporter on LightCycler Using Hybridization Probes and External Standards

Danuta Kielar, Wolfgang Dietmaier, Thomas Langmann,
Charalampos Aslanidis, Mario Probst, Marek Naruszewicz,
Gerd Schmitz*

Introduction

ABCA1 is a member of the subfamily A of ATP-binding cassette transporters, whose recently discovered mutations cause familial HDL deficiency syndromes such as Tangier disease [1–3]. Tangier patients have very low HDL levels, associated either with the classical phenotype (splenomegaly) or with premature atherosclerosis and a higher risk of developing coronary artery disease. Overexpression studies and analysis of *ABCA1* knock-out mice have demonstrated that ABCA1 plays a key role in cellular transport of cholesterol and phospholipids [4, 5]. ABCA1 mRNA and protein in human macrophages are up-regulated by cholesterol loading and down-regulated by cholesterol deloading. Mechanisms controlling *ABCA1* gene expression comprise sterol-dependent and tissue-specific pathways. *ABCA1* gene expression is up-regulated by modified LDL [6] cyclic AMP [7, 8] and oxysterols, which act on LXRα and LXRβ receptors together with the retinoid X receptor (RXR) [9, 10]. A zinc finger transcription factor (ZNF202) located within a hypoalphalipoproteinemia locus on chromosome *11q23* has a strong repressive capacity on *ABCA1* expression as well as cellular lipid efflux [11]. However, the influence of other metabolic factors on the expression of *ABCA1* in humans is still not well understood. Therefore, a rapid, sensitive, specific and reproducible method for detection and quantification of ABCA1 transcripts is needed. In this chapter, we describe a real-time RT-PCR technique for detection and quantification of minute amounts of ABCA1 mRNA. Using this method we quantified ABCA1 transcripts in various human tissues, as well as in monocytes, macrophages, THP-1 cells, fibroblasts, adipocytes, preadipocytes, keratinocytes and in HeLa cells transfected with Sp1 and Sp3 transcription factors.

---

* Gerd Schmitz (✉) (e-mail: gerd.schmitz@klinik.uni-regensburg.de)
  Institute for Clinical Chemistry and Laboratory Medicine, University of Regensburg,
  Franz-Josef-Strauss-Allee 11, 93042 Regensburg, Germany
  Wolfgang Dietmaier
  Institute of Pathology University of Regensburg, 93042 Regensburg, Germany
  Danuta Kielar, Marek Naruszewicz
  Department of Clinical Biochemistry, Pomeranian Medical Academy, 70–111 Szczecin, Poland

Materials

Equipment
LightCycler Instrument (Roche Molecular Biochemicals, Mannheim, Germany)
LightCycler Capillaries (Roche Molecular Biochemicals)
ABI Prism Genetic Analyzer 310 (PE Biosystem, ForsterCity, USA)

Kits
FastStart DNA Master Hybridization Probes Kit (Roche Molecular Biochemicals)
RNA 6000 LabChip Kit (Agilent Technologies, Böblingen, Germany)
TOPO TA Cloning, (Invitrogen, Groningen, The Netherlands)
Qiaprep Spin Miniprep Kit (QIAGEN, Hilden, Germany)
RiboProbe In Vitro Transcription System (Promega, Mannheim, Germany)
1st Strand cDNA Synthesis Kit for RT-PCR (AMV), (Roche Molecular Biochemicals)

Reagents
Amplification Primers (Metabion, Munich, Germany)
Hybridization Probes (TIB MOLBIOL, Berlin, Germany)
DNase I (Roche Diagnostics, Mannheim, Germany)
Trizol reagent (GIBCO BRL, Eggenstein, Germany)
*Bam*HI restriction enzyme (Roche Molecular Biochemicals)

Procedure

Sample Preparation
Expression analysis of ABCA1 transcripts on the LightCycler instrument was performed using different cell types grown by standard culture methods. The following in vitro cultured cells were used: THP-1 cells (obtained from ATCC), HeLa cells, keratinocytes (NHEK-Normal Human Epidermal Keratinocytes), HaCaT cells (spontaneously immortalized human keratinocyte cell line) and fibroblasts. Monocytes were obtained from healthy normolipidemic volunteers by leukapheresis and were purified by counterflow elutriation and were further differentiated into macrophages in the presence of human recombinant M-CSF. The basal expression of ABCA1 mRNA, the effect of differentiation (adipocyte, monocyte, keratinocyte, HaCaT), effects of lipid loading and deloading (macrophage, THP-1 cells, fibroblasts) and its regulation by transcription factors (experiments with Sp1- and Sp3-transfected HeLa cells) were studied. Total RNA from all cultured cells was isolated with Trizol reagent (Sigma). All RNA samples were treated with DNase I (Roche Diagnostics) according to the protocol of Huang et al. [12]. Total RNA from different human tissues was obtained from Clontech (Heidelberg, Germany). Human subcutaneous preadipocytes and adipocytes were kindly provided by Prof. Georg Löffler (Institute for Biochemistry, Genetics and Microbiology, University of Regensburg, Germany).

The concentration, purity and integrity of total RNA from all samples was assessed using the Agilent 2100 bioanalyzer and the RNA 6000 LabChip Kit (Agilent Technologies). Briefly, the procedure is based on labelling of the RNA with a fluorescent dye and, upon electrophoretic separation in capillaries embedded in the chip, the 18S and 28S RNA peaks are identified. The presence of very distinct and sharp peaks for 18S and 28S and a ratio of ~2 (28S/18S) is indicative for high quality, non-degraded RNA. Concentration is calculated from the fluorescence signal obtained. Only high-quality RNA was used.

ABCA1-specific PCR primers and hybridization probes, capable of fluorescence resonance energy transfer (FRET) were used to generate and monitor a 205-bp ABCA1 fragment and a 282-bp porphobilinogen deaminase fragment (*PBGD*), respectively (Table 1). Primers and hybridization probes for ABCA1 and *PBGD* were synthesized by TIB MOLBIOL. Hybridization probes were labeled with fluorescein at the 3′ end of the donor probe and LCRed640 at the 5′ end of the acceptor probe.

Oligonucleotide Design

A 205-bp ABCA1 RT-PCR product was subcloned in the plasmid vector pCR[R] II-TOPO containing a T7 RNA polymerase promoter (TOPO TA Cloning, Invitrogen) according to the manufacturer's instructions. Isolation and purification of plasmid DNA was performed using the Qiaprep Spin Miniprep Kit (QIAGEN, Germany). The *ABCA1* insert sequence was verified by DNA sequencing using an ABI Prism Genetic Analyzer 310 (PE Biosystem). Linearization of purified plasmid DNA was achieved by incubation with *Bam*HI and RNA was synthesized in vitro using the RiboProbe[R] In Vitro Transcription System (Promega) and T7 RNA Polymerase following the protocol for large scale RNA synthesis.

Generation of an External ABCA1 RNA Standard

First-strand cDNA synthesis was performed in a total volume of 20 µl using 40 U AMV Reverse Transcriptase, 2.0 µl 10× Reaction Buffer, 50 U RNase Inhibitor, 2 µl Deoxynucleotide Mix (1 mM), 2 µl Random Primer p(dN)$_6$ (3.2 µg), 4 µl MgCl$_2$

Reverse Transcription

Table 1. Oligonucleotides

GenBank Accession #AB055982 (human ABCA1 mRNA), X04217 (human PBGD mRNA)				
	Position	Length	GC (%)	T_m (°C)
ABCA1 forward primer GCACTGAGGAAGATGCTGAAA	1013	21	47.6	60,3
ABCA1 reverse primer AGTTCCTGGAAGGTCTTGTTCAC	1217R	23	47.8	60,8
ABCA1 donor probe GCCGCTGCTCGTTGGGAAGAT-F	1128	21	61.9	70,1
ABCA1 acceptor probe LCRed640-CTGTATACACCTGACACTCCAG-P	1150	22	50.0	54,4
PBGD forward primer: AGAGTGATTCGCGTGGGTACC	85	21	57.1	63,5
PBGD reverse primer: GGCTCCGATGGTGAAGCC	366	18	66.7	64,0
PBGD donor probe: AGTGGACCTGGTTGTTGCACTCCTTGAA-F	297	28	48.1	72,3
PBGD acceptor probe: LCRed640-ACCTGCCCACTGTGCTTCCTCCT-P	326	23	60.9	69,7

(5 mM), and 1 µg total RNA, as recommended by the manufacturer [1st Strand cDNA Synthesis Kit for RT-PCR (AMV), Roche Molecular Biochemicals]. Serial dilutions of in vitro synthesized *ABCA1* RNA (100 pg, 10 pg, 1 pg, 0.1 pg) were reverse transcribed and 1/10 of each reaction was used for generation of a standard curve (Fig. 2). Following cDNA synthesis for 60 min at 42°C and inactivation of the enzyme at 95°C for 5 min, PCR reactions were performed in the LightCycler instrument.

LightCycler PCR The following Hybridization Probe Master Mix was used for amplification of the *ABCA1*- and *PBGD*-specific fragments. The same reaction mix was used for amplification of the ABCA1 RNA standard.

	Volume [µl]	[Final]
Hot Start PCR Reaction Mix	1.5	1×
$MgCl_2$ (25 mM)	2.4	5.0 mM
Primers (25 µM each)	0.3+0.3	0.5 µM
Probes (20 µM each)	0.22+0.22	0.3 µM
H_2O (PCR-grade)	8.06	
Total volume	13	

We added 13 µl of master mix and 2 µl of RT reaction mixture (usually corresponding to 100 ng of total RNA) to each glass capillary. All of the reverse transcribed standard dilutions were amplified simultaneously. Sealed capillaries were centrifuged briefly with the adapters in a microcentrifuge and put in the Light-Cycler carousel.

The following PCR protocol was used for amplification and melting curve analysis:
- Denaturation

Parameter	Value
Cycles	1
Type	Regular Segment 1
Target temperature [°C]	95
Incubation time [min]	10
Temperature transition rate [°C/s]	20
Acquisition Mode	None

- Amplification

Parameter	Value		
Cycles	45		
Type	Quantification		
	Segment 1	Segment 2	Segment 3
Target temperature [°C]	95	60	72
Incubation time [s]	10	10	5
Temperature transition rate [°C/s]	20	20	20
Acquisition mode	None	Single	None
Gains	F1=5; F2=10		

- Melting Curve Analysis

Parameter	Value		
Cycles	1		
Type	Melting curves		
	Segment 1	Segment 2	Segment 3
Target temperature [°C]	95	50	95
Incubation time [s]	0	10	0
Temperature transition rate [°C/s]	20	20	0.2
Acquisition mode	None	None	Step
Gains	F1=5; F2=10		

- Cooling was for 30 s at 40°C (temperature transition rate, 20°C/s)

Results

To amplify a human *ABCA1*-specific 205-bp RT-PCR fragment, PCR primers were located in exon 8 and exon 10. As shown in Fig. 1a, these primers span across introns 8 and 9, thereby preventing the amplification of genomic DNA templates. The hybridization probes are separated by two nucleotides and are complementary to the sequence within the central region of exon 9. The normalization of ABCA1 mRNA expression levels was achieved using specific primers and hybridization probes for a human porphobilinogen deaminase (PBGD) 282-bp fragment (Fig. 1b). Primers and hybridization probes sequences are shown in Table 1. To achieve absolute quantification of ABCA1 mRNA, serially diluted in vitro transcribed ABCA1 RNA was used as an external control. Four dilutions of ABCA1 cRNA were reverse transcribed in parallel to the sample material to be analysed. The standard curve (Fig. 2b) was used for quantification of both the ABCA1 transcript and the housekeeping gene (*PBGD*) when relative quantification was calculated. The PBGD was amplified with the same efficiency as *ABCA1* mRNA (data not shown). One aliquot of cDNA was used to analyze ABCA1 in

Fig. 1a,b. Positions of PCR primers and hybridization probes used in the assay. **a** *ABCA1* fragment. A pair of primers (*dotted arrows, italics*) amplify a 205-bp fragment of the ABCA1 cDNA, containing a part of exon 8, exon 9 and the first 23 bp of exon 10 (position in mRNA: 1013–1217 bp). The ABCA1 donor hybridization probe (*green solid line*) and ABCA1 acceptor hybridization probe (*red solid line*) are separated by one nucleotide. **b** PBGD fragment. A pair of primers (*dotted arrows, italics*) amplify a 282-bp fragment of PBGD cDNA, containing a part of exon 2, exon 3, exon 4 and almost the entire exon 6 (position mRNA 85–366 bp). The PBGD donor hybridization probe (*green solid line*) and the PBGD acceptor hybridization probe (*red solid line*) are separated by two nucleotides

each sample, whereas another aliquot served to determine PBGD expression. In Fig. 2a, the real-time PCR curves from the amplification of the ABCA1 cRNA dilutions (for generating the standard curve) and unknown sample (lung in this example) are displayed. The amplification of both transcripts was performed simultaneously during the same run. Figure 3a and b (top panels) displays the amplification of *ABCA1* and *PBGD* targets in various human tissues. To characterize both amplified products, melting curves were generated for all samples

Amplification

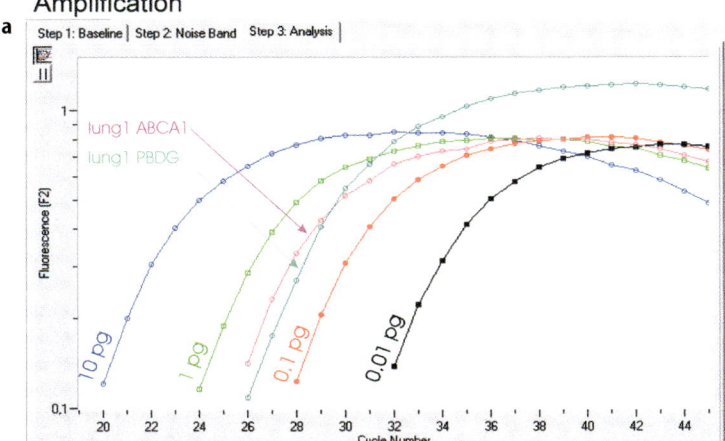

Standard

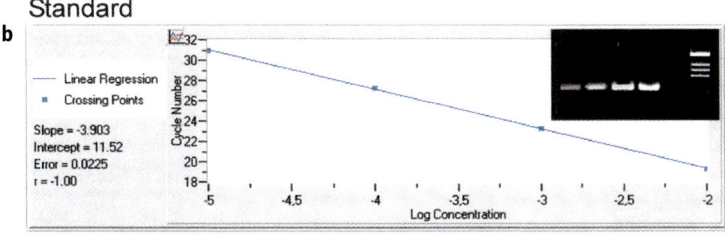

Melting Curves

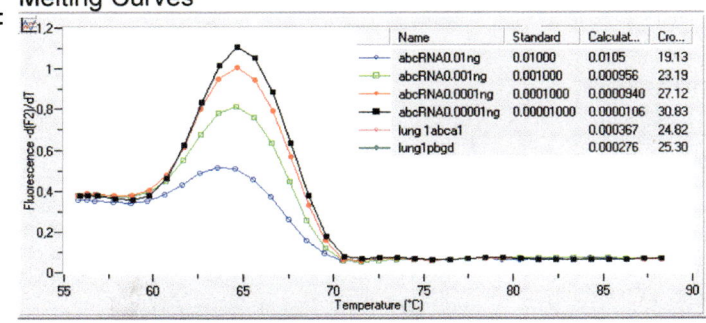

Fig. 2a,b,c. Detection and quantification of ABCA1 product in the LightCycler using the Hybridization Probe Format and standard curve. Serial dilutions (10 pg, 1 pg, 0.1 pg) of in vitro transcribed ABCA1 cRNA fragment (external standard) were first reverse transcribed and 1/10 of each RT reaction mixture (10 pg, 1 pg, 0.1 pg, 0.01 pg) was amplified (**a**) to generate a standard curve for absolute ABCA1 and PBGD transcript quantification (**b**) and a melting curve analysis (**c**). The Y intercept =11.52, equivalent to the log of the amount of PCR product at the crossing point-Cp divided by the log of the efficiency of the reaction E; $E=10^{-1/slope} =10^{-1/-3.903}=1.80$; slope =−3.903, equivalent to −1/log E; error =0.0225 (mean squared error), $r^2=1$ (regression coefficient). The Cp values of the unknown sample are converted to concentrations, using the equation derived from the standard. Gel analysis of standard amplifications in 2% agarose is displayed (**b**). Only specific products of each dilution at approximately 200 bp are visible. (M:1-kb marker)

ABCA1 Amplification

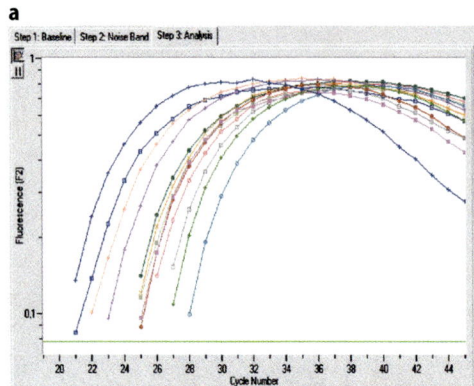

PBGD1 Amplification

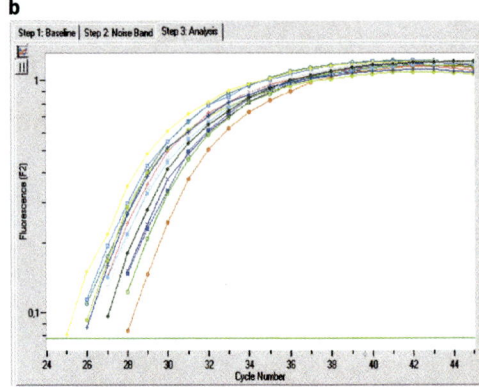

ABCA1 Melting Curves

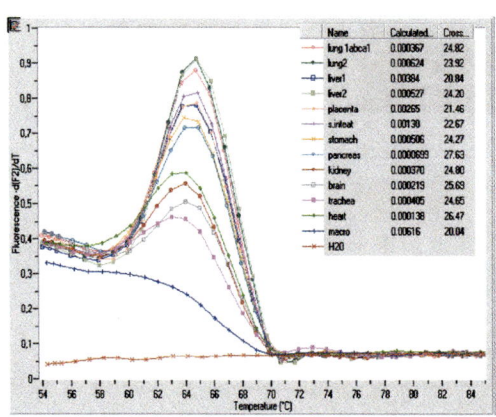

PBGD Melting Curves

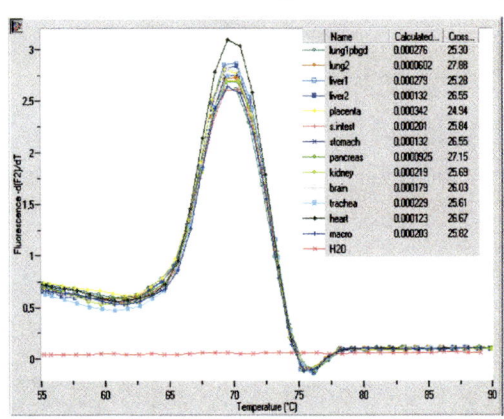

Tissue expression (absolute)

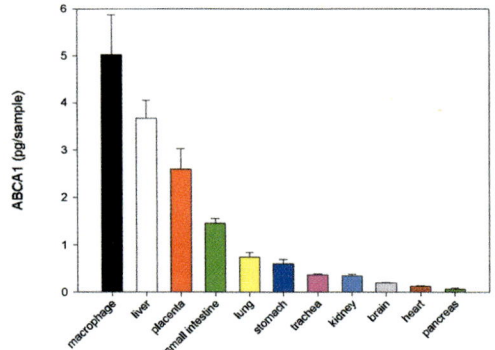

Tissue expression (ratio)

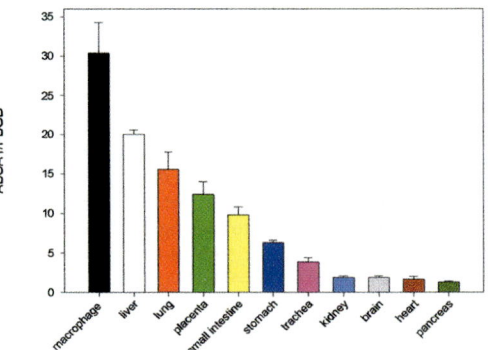

after each run and then converted to derivative melting curves (Fig. 3a,b, middle panels). The melting points (or T_m, melting temperatures) serve to identify and discriminate between both analyzed products (65°C and 70°C for *ABCA1* and *PBGD*, respectively).

The ABCA1 mRNA expression levels of various tissues were compared to macrophages. The results are presented either as an absolute amount of *ABCA1* mRNA based on a calculation using the standard curve (pg/sample) or are standardized by the quantification of *PBGD* (Fig. 3a,b, respectively, bottom panel). From all samples examined, the highest levels of ABCA1 mRNA were found in macrophages, liver, and stomach. The lowest amounts of ABCA1 transcripts were detected in the pancreas and heart. Using our assay we have demonstrated ABCA1 mRNA up-regulation during adipocyte (2.5-fold increase) and monocyte differentiation (1.5-fold increase). We have found it very useful in exploring the effect of cholesterol loading and deloading on *ABCA1* expression in macrophages and THP-1 cells. In addition, we were able to detect ABCA1 mRNA in keratinocytes, HaCaT cells and in stem cells isolated from cord blood, which previously was impossible with conventional RT-PCR. With our quantitative RT-PCR we investigated the effect of Sp1 and Sp3 transcription factors on ABCA1 mRNA expression in HeLa cells transfected. We found a dose-dependent up-regulation of ABCA1 transcript by Sp1.

Comments

Quantification of the absolute *ABCA1* gene transcript number is a very important issue in many studies related to its expression and regulation by metabolic factors, nutritional status and new antiatherogenic drug candidates. In the majority of publications investigating *ABCA1* expression, conventional methods for gene expression analysis were used, mainly Northern blot, a time-consuming and relatively less sensitive method that is difficult to standardize. Here we demonstrate that a LightCycler-based ABCA1 mRNA assay allows the amplification, detection and quantification of minute amounts of ABCA1 mRNA in a variety of tissues and cells. Our hybridization probe assay not only makes it possible to estimate the absolute amounts of *ABCA1* transcripts, but also allows determination of the rel-

◄

Fig. 3a,b. Amplification, quantification and melting peak analysis of ABCA1 (**a**) and PBGD (**b**) PCR products from human tissues. Of total RNA, 1 μg from each tissue was reverse transcribed and 2 μl aliquots of RT reaction mixture were amplified in the LightCycler in the presence of hybridization probes. Melting curves were converted to derivative melting curves by plotting the first negative derivative of the fluorescence with respect to the temperature (–dF/dT) against the temperature. Melting temperatures (T_m) identify each of the products (ABCA1 T_m =65°C; PBGD T_m =70°C). No run-to-run variation in T_m values was observed. The absolute amounts (pg/sample) of ABCA1 transcripts were calculated as described in Fig. 2. The ABCA1/PBGD ratios were calculated by dividing absolute amounts of ABCA1 and PBGD determined for each sample. *Bars* represent mean values (±SD) of four independent measurements: **a**, absolute amount of *ABCA1* transcript (*bottom*); **b**, ABCA1/PBGD ratio (*bottom*)

ative changes in its expression (ABCA1/PBGD ratio). The advantage of using the hybridization probe format is its specificity, as confirmed by melting curve analysis for each amplified product, ABCA1 and PBGD in our case. Melting curve analysis at the end of the LightCycler run enables the characterization of the amplified product and obviates gel electrophoresis. Our data show that standardization with a housekeeping gene (e.g., *PBGD*) is not completely satisfactory when the target gene (e.g., *ABCA1*) is expressed in different tissues and different cells. Because *PBGD* is not equally expressed in various tissues, absolute quantification may be more reliable. However, when we normalized our absolute quantification of ABCA1 mRNA levels with *PBGD* as a reference gene in the same cell type (e.g., fibroblasts, monocytes, HeLa cells) we obtained comparable results. In contrast, the other commonly used internal standard glyceraldehyde-3-phosphate-dehydrogenase (GAPDH) varies quantitatively in response to various factors making the interpretation difficult [13, 14]. Based on initial experiments with GAPDH as an endogenous control, we decided to use porphobilinogen deaminase (*PBGD*) as the reference gene [15]. Unlike GAPDH, we have amplified both ABCA1 and PBGD transcripts with the same efficiency, which is an important factor for the final calculation. In contrast to GAPDH, the human *PBGD* gene is free of pseudogenes [16] and only cDNA-derived products are accumulated during PCR, so that DNase digestion is not required. We recommend *PBGD* as a normalization control for mononuclear cells, fibroblasts, adipocytes, HepG2 cells, HeLa cells and THP-1 cells. The two-step quantitative RT-PCR with external standards and hybridization probes is rapid, reliable, reproducible and not as laborious as other technologies. The high sensitivity of this method allowed detection and quantification of ABCA1 mRNA in a variety of human cells and tissues.

Applications This quantitative method can be used: (a) in the monitoring of drug effects, (b) in epidemiological studies, (c) in studies assessing correlation between *ABCA1* expression and susceptibility to lipid disorders, and (d) in exploring the effects of polymorphisms in the promoter region of the *ABCA1* gene.

References

1. Bodzioch M, Orso E, Klucken J, Langmann T, Böttcher A, Diederich W, Drobnik W, Barlage S, Buchler C, Porch-Ozcurumez M, Kaminski WE, Hahmann HW, Oette K, Rothe G, Aslanidis C, Lackner K, Schmitz G (1999) The gene encoding ATP-binding cassette transporter 1 is mutated in Tangier disease. Nat Genet 22:347–351
2. Brooks-Wilson A, Marcil M, Clee S, Zhang LH, Roomp K, vanDam M, Yu L, Brewer C, Collins JA, Molhuizen HO, Loubser O, Ouelette BF, Fichter K, Ashobourne-Excoffon KJ, Sensen CW, Scherer S, Mott S, Denis M, Martindale D, Frohlich J, Morgan K, Koop B, Pimstone S, Kastelein JJ, Hayden MR (1999) Mutations in ABC1 in Tangier disease and familial high-density lipoprotein deficiency. Nat Genet 22:336–345
3. Rust S, Rosier M, Funke H, Real J, Amoura Z, Piette JC, Deleuze JF, Brewer HB, Duverger N, Denefle P, Assman G (1999) Tangier disease is caused by mutations in the gene encoding ATP-binding cassette transporter 1. Nat Genet 22:352–355
4. Orso E, Broccardo C, Kaminski WE, Böttcher A, Liebisch G, Drobnik W, Götz A, Chambenoit O, Diederich W, Langmann T, Spruss T, Luciani MF, Rothe G, Lackner KJ, Chimini G, Schmitz

G (2000). Transport of lipids from golgi to plasma membrane is defective in Tangier disease patients and Abc1-deficient mice. Nat Genet 24:192–196
5. Lawn RM, Wade DP, Garvin MR, Wang X, Schwartz K, Porter JG, Seilhamer JJ, Vaughan MV, Oram JF (1999) The Tangier disease gene product ABC1 controls the cellular apolipoprotein-mediated lipid removal pathway. J Clin Invest 104:R25–R31
6. Langmann T, Klucken J, Reil M, Liebisch G, Luciani MF, Chimini G, Kaminski WE, Schmitz G (1999) Molecular cloning of the human ATP-binding cassette transporter 1 (hABC1): evidence for sterol-dependent regulation in macrophages. Biochem Biophys Res Commun 257:29–33
7. Oram JF, Lawn RM, Garver MR, Wade DP (2000) ABCA1 is the cAMP-inducible receptor that mediates cholesterol secretion from macrophages. J Biol Chem 275:34508–34511
8. Abe-Dohmae S, Suzuki S, Wada Y, Aburatani H, Vance DE, Yokoyama SR (2000) Characterization of apolipoprotein-mediated HDL generation induced by cAMP in a murine macrophage cell line. Biochemistry 39:11092–11099
9. Costet P, Luo Y, Wang N, Tall AR (2000) Sterol-dependent transactivation of the ABC1 promoter by the liver X receptor/retinoid X receptor. J Biol Chem 275:28240–28245
10. Repa JJ, Turley SD, Lobaccaro JA, Medina J, Li L, Lustig K, Shan B, Heyman RA, Dietschy JM, Mangelsdorf DJ (2000) Regulation of absorption and ABC1-mediated efflux of cholesterol by RXR heterodimers. Science 289:1524–1529
11. Porsch-Özcürümez M, Langmann T, Heimerl S, Borsukova H, Kaminski WE, Drobnik W, Honer C, Schumecher C, Schmitz G (2001) The zinc finger protein (ZNF202) is a transcriptional repressor of ABCA1 and ABCG1 gene expression and a modulator of cellular lipid efflux. J Biol Chem 276 (15):12427–12433
12. Huang Z, Fasco MJ, Kaminsky LS (1996) Optimisation of Dnase I removal of contaminating DNA from RNA for use in quantitative RNA-PCR. Biotechnique 20:1012–1020
13. Bustin SA (2000) Absolute quantification of mRNA using real-time reverse transcription polymerase chain reaction assays. J Mol Endocrinol 25:169–193
14. Thellin O, Zorzi W, Lakaye B, De Borman B, Coumans B, Hennen G, Grisar T, Igout A, Heinen E. (1999) Housekeeping genes as internal standards: use and limits. J Biotechnol 75:291–295
15. Nagel S, Schmidt M, Thiede C, Huhn D, Neubauer A (1996) Quantification of Bcr-Abl transcripts in chronic myelogenous leukemia (CML) using standardized, internally controlled, competitive differential PCR (CD-PCR). Nucleic Acids Res 24:4102–4103
16. Finke J, Fritzen R, Ternes P, Lange W, Dölken G (1993) An improved strategy and useful housekeeping gene for RNA analysis from formalin-fixed, paraffin-embedded tissues by PCR. Biotechniques 14:448–453

Quantification of Human Genomic DNA Using Retinoic X Receptor B Gene

NATHALIE PIERI-BALANDRAUD, JEAN ROUDIER, CHANTAL ROUDIER*

Introduction

The LightCycler** system performs real-time PCR with specific and quantitative analysis of amplified products. All parameters of the PCR reaction have to be carefully adjusted. The most important factors are DNA quality and concentration. Pure DNA is prepared with any manufactured kit. Quantification of genomic DNA samples is essential for accurate quantification of small differences in gene copy numbers or low copy target genes. Classic methods used to quantify genomic DNA have limitations.

The traditional method to determine the amount of DNA in solution by measuring ultraviolet absorbance at 260 nm is easy to perform but has serious limitations [1]; for example, absorbance readings cannot discriminate between DNA and RNA. The absorbance ratio at 260 and 280 nm is used as an indicator of nucleic acid purity. Proteins have a peak absorption at 280 nm and reduce the A260/A280 ratio. Ratios of 1.8 to 2 indicate highly purified preparations of DNA or RNA. Absorbances at 230 nm and 325 nm reflect contamination by phenol, urea and particulates, and dirty cuvettes, respectively. Absorbance at 320 nm should be approximately zero with pure samples. Therefore, careful handling of sample cuvettes is recommended. A large amount of DNA is required to obtain accurate readings. An absorbance of 1.0 at 260 nm indicates 50 µg/ml of double-stranded DNA.

Ethidium bromide fluorescent quantification of doubled-stranded DNA is also widely used [1]. Electrophoresis through gels containing ethidium bromide (0.5µg/ml) is carried out for this purpose. Known volumes of DNA samples and a series of standard DNA solutions (500, 250, 125, 63, 31 and 15 ng/6 µl) are loaded on the same gel. After migration the gel is photographed using short-wavelength ultraviolet radiation. The intensity of fluorescence of the unknown DNA is compared with that of the DNA standards and the quantity of DNA in the sample is estimated. A critical parameter is the "visual reading" of fluorescence.

We developed a simple technique to quantify DNA using the LightCycler System. Retinoic X Receptor B is a human single-copy gene mapping to chromosome

---

* Chantal Roudier (✉) (email: chantal.roudier@medecine.univ-mrs.fr)
 Laboratoire Immunorhumatologie, INSERM EMI 9940, Faculté de Medecine,
 27 bd Jean Moulin, 13005 Marseille, France
** LightCycler is a trademark of a member of the Roche Group

6p21.3-p21.1 [2]. A 150-bp segment of intron 1 is amplified in standard DNA and in test DNA in the same conditions. This amplification precisely reflects the number of genomic copies present in the sample.

Materials

LightCycler instrument and software (Roche Diagnostics, Mannheim, Germany)
LightCycler capillaries, centrifuge adaptors and cooling blocks (Roche Diagnostics)
Human Blood (from the lab staff)
Genomic DNA (Clontech, Palo Alto, USA)
LightCycler FastStart DNA Master SYBR Green I (Roche Diagnostics)
LightCycler FastStart DNA Master Hybridization Probes (Roche Diagnostics)
QIAGEN Genomic Tips (100/G) (Qiagen, Hilden, Germany)
QIAGEN Genomic DNA Buffer Set
QIAGEN Protease
Retinoic X Receptor B primers and probes designed, synthesized, and purified (HPLC) by TIB MOLBIOL (Berlin, Germany)

Procedure

Sample Preparation

Human genomic DNA was isolated from 5 ml of heparinized blood. Mononuclear cells were isolated by isopycnic centrifugation through Ficoll-Histopaque (Sigma St Louis, USA) and processed according to QIAGEN genomic DNA purification Handbook. DNA was resuspended in 10 mM Tris, pH 8. DNA concentration was estimated by the ethidium bromide fluorescence technique and DNA quality was evaluated with ultra violet absorbance. Stock sample DNA solutions (approximately 500–100 ng/µl) were diluted to 1:10 and 1:100.

A dilution series in the range of 100–0.01 ng/µl of Clontech genomic DNA was used for external standardization. Since the mass of one human haploid genome is 3 pg, these dilutions correspond approximately to 33,333–3.3 copies of the single-copy *RXRB* gene.

Oligonucleotides are shown in Table 1.

Table 1. Oligonucleotides

Homo sapiens Retinoic X Receptor B gene (GenBank Accession #AF065396)				
	Position	Length	GC (%)	T_m (°C)
PCR primers				
GAG CGA CGG GCT TAA TTC GA	714	20	55	60.3
GAG CGG CCC AAG ACA TG	864	17	64.7	57.5
PCR Product		150	56.7	86.7
Hybridization probes				
TCG GAG GAT TAG CTG AGC ACG AGG A-X	792	25	56.0	65.7
LCRed640-CCC CTG AGA GAA AGA CTC TGG CCT G-P	765	25	60.0	66.2

Preparation of SYBR Green I master mix and hybridization probes master mix:

	Volume [µl]		[Final]
	SYBR Green I	Hybridization	
H_2O	10.4	8.4	
$MgCl_2$ (25 mM/l)	1.6	1.6	3 mM
Primers (10 µM each)	1+1	1+1	0.5 µM each
LightCycler FaststStart DNA Master SYBR Green I	2		1×
LightCycler Hybridization Probe Master		2	1×
Hybridization probes (3 µM each)		1+1	0.15 µM each
Total master mix volume per reaction	16	16	20 µl final volume

For n PCR reactions, reagents for $n+1$ reactions were mixed and 16 µl pipetted into each precooled capillary. To reduce pipetting error, 4 µl of standard or test DNA template were then added.

Protocol for amplification with SYBR Green I:

Parameter	Program 1	Program 2			Program 3		
Cycles	1	40			1		
Type	Denaturation	Quantification			Melting Curve		
	Seg 1	Seg 1	Seg 2	Seg 3	Seg 1	Seg 2	Seg 3
Target temperature [°C]	95	95	60	72	95	70	40
Incubation time [s]	600	15	5	10	30	0	50
Temperature transition rate [°C/s]	20	20	20	20	20	0.1	20
Acquisition mode	None	None	None	Single	Cont	Cont	Cont
Gains	F1=4	F1/1					

Protocol for amplification with hybridization probes:

Parameter	Program 1	Program 2			Program 3
Cycles	1	40			1
Type	Denaturation	Quantification			Cooling
	Seg 1	Seg 1	Seg 2	Seg 3	Seg 1
Target temperature [°C]	95	95	60	72	40
Incubation time [s]	600	15	10	10	30
Temperature transition rate [°C/s]	20	20	20	5	20
Acquisition mode	None	None	Single	None	None
Gains	F1=5; F2=15				

Results

For SYBR Green I experiments, typical melting curves were obtained with a product Tm of 86.7°C (Fig. 1). Very few primer-dimers were present, even with the lowest concentration of standard DNA, negative control or test DNA. We verified the absence of non-specific PCR products and primer-dimers on a 2% agarose gel.

A classic calibration curve was obtained with a twofold dilution series of DNA. Data were analyzed using the fit points method (2-points). Standard curves obtained with SYBR Green I (Fig. 2) and hybridization probes (Fig. 3) were linear over at least 4 orders of magnitude. When needed, concentrations of genomic DNA were adjusted by appropriate dilution to enter the standard range. Genomic DNA was then quantified by comparison to the standard concentration curve.

Comments

The purpose of this work is to quantify human genomic DNA by real-time PCR using an external DNA concentration standard. Clearly, gene quantification normalized to the amount of DNA prepared from a certain volume of blood is inadequate, even when precision and reproducibility of the technique are emphasized. Although the concentration of the genomic DNA standard may not be accurate, if all samples are compared to a single standard, the relative values for all samples will be correct. New standards should be compared to known standards (Fig. 4).

DNA quantification was performed on ethidium bromide stained gel and compared to DNA quantification with LightCycler hybridization probes. Although similar values were obtained with both techniques, the precision of the LightCycler data was much better than gel quantification (Fig.5).

The LightCycler technique makes it possible to quantify small amounts of genomic DNA (0.04–400 ng) precisely. Absorbance and ethidium bromide techniques are restricted to high DNA concentrations and provide approximate estimation of DNA concentration.

Application

Amplification of Retinoic X Receptor B gene with the LightCycler provides specific and precise quantification of human DNA and can be applied to any kind of human DNA sample. Quantification of other genes is then related to a precise DNA concentration or a precise number of genome copies.

We are currently quantifying low copy viral genes in human lymphocytes. Therefore, we need accurate quantification of lymphocyte DNA. We perform such quantification using *RXRB* amplification. We then perform viral DNA amplification from a well known quantity of lymphocyte DNA. Coamplification of *RXRB* and the gene of interest in a single tube should give even more accurate quantification. In a recent article, Jabs et al. described such coamplification of the genomic C-reactive protein gene and Epstein Barr virus [3].

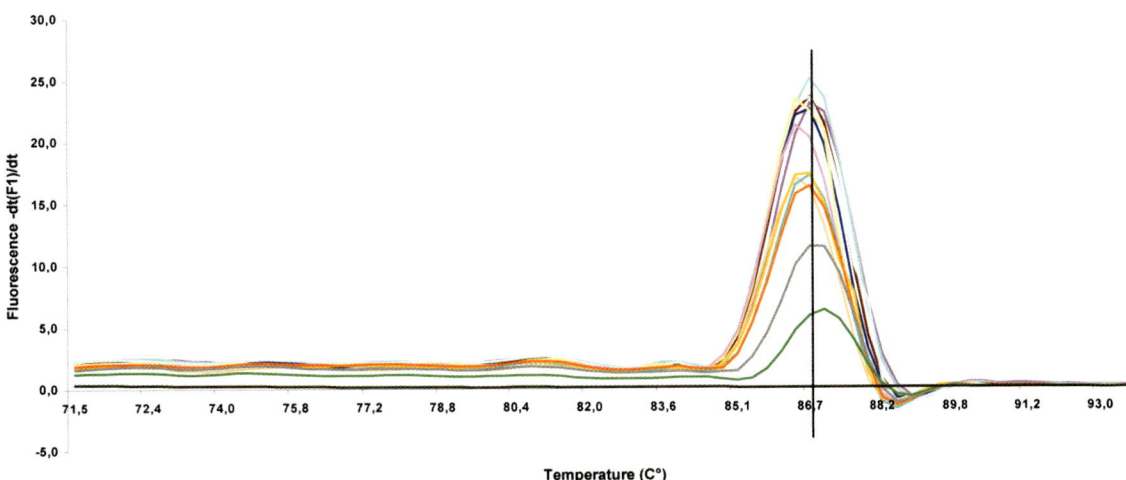

Fig. 1. SYBR Green I melting curve of the amplified retinoic X receptor B gene. The negative derivative of fluorescence versus temperature is shown. All DNA dilutions show the same profile with an average T_m of 86.7

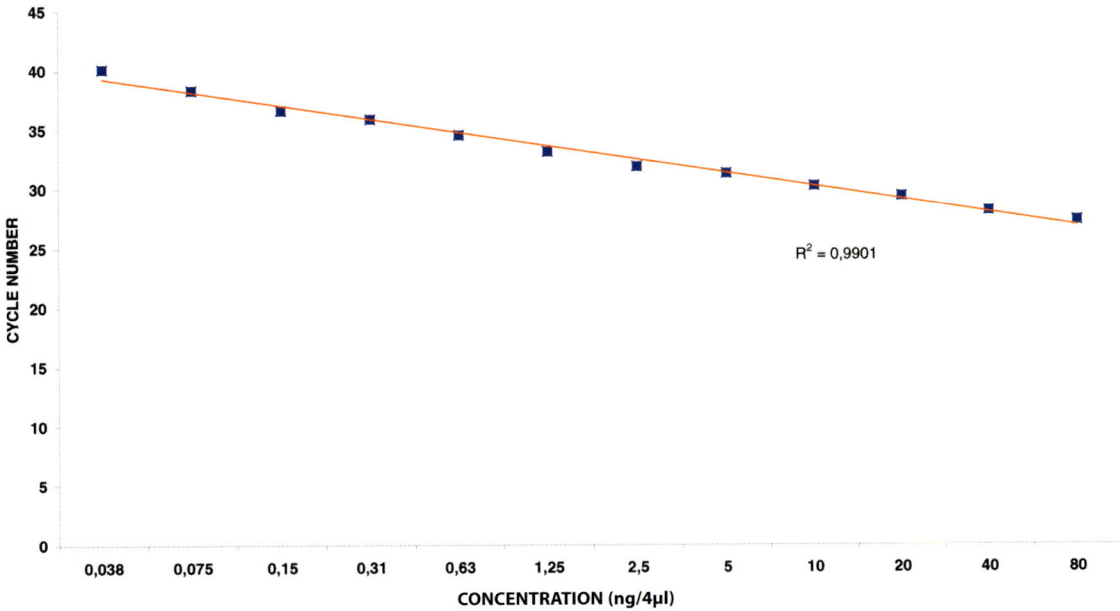

Fig. 2. Standard curve using SYBR Green I. A dilution series of human genomic DNA was used as standard DNA template for amplification of a *RXRB* 150-bp fragment. The calibration curve shows the logarithmic plot of the *RXRB* template concentration versus cycle number

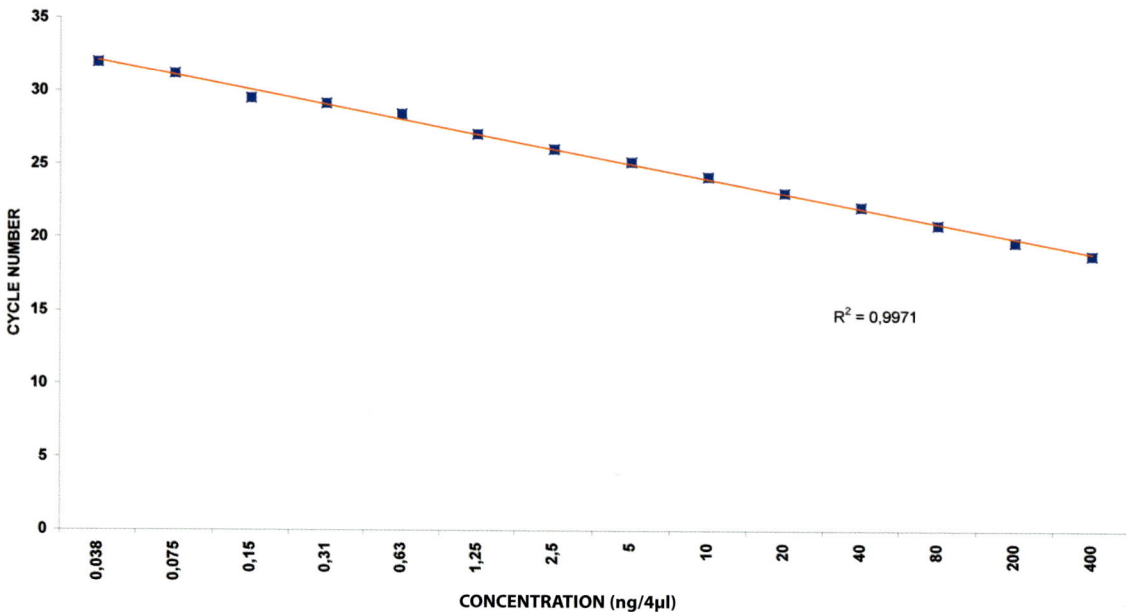

Fig. 3. Standard curve using hybridization probes. Conditions were similar to Fig. 2

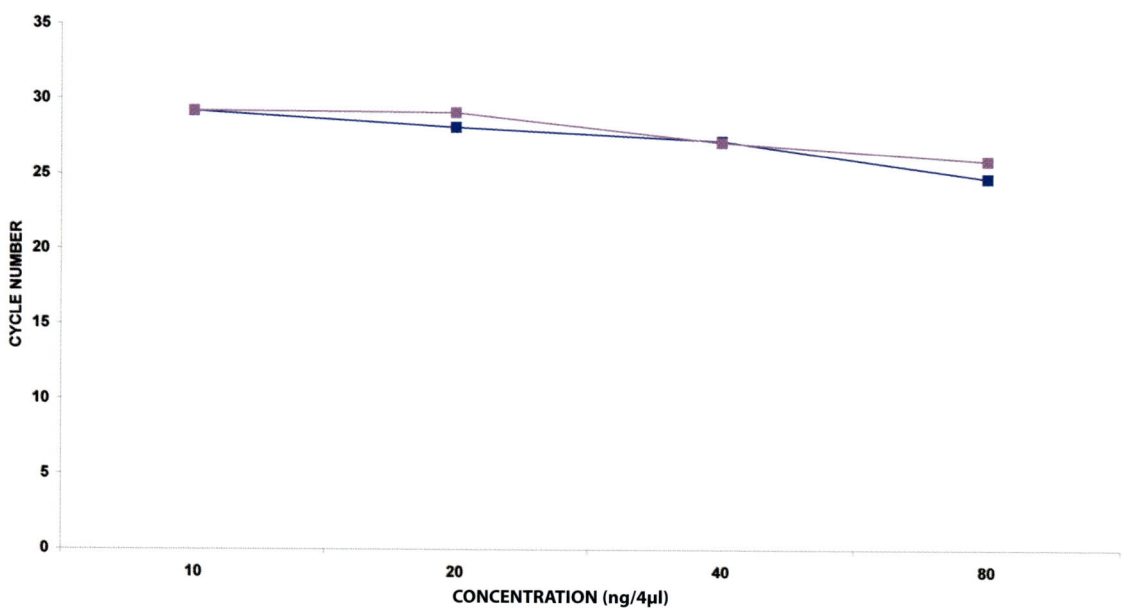

Fig. 4. Comparison of standards. Four new standards dilutions (*pink*) are plotted against identical dilutions of a prior standard (*blue*)

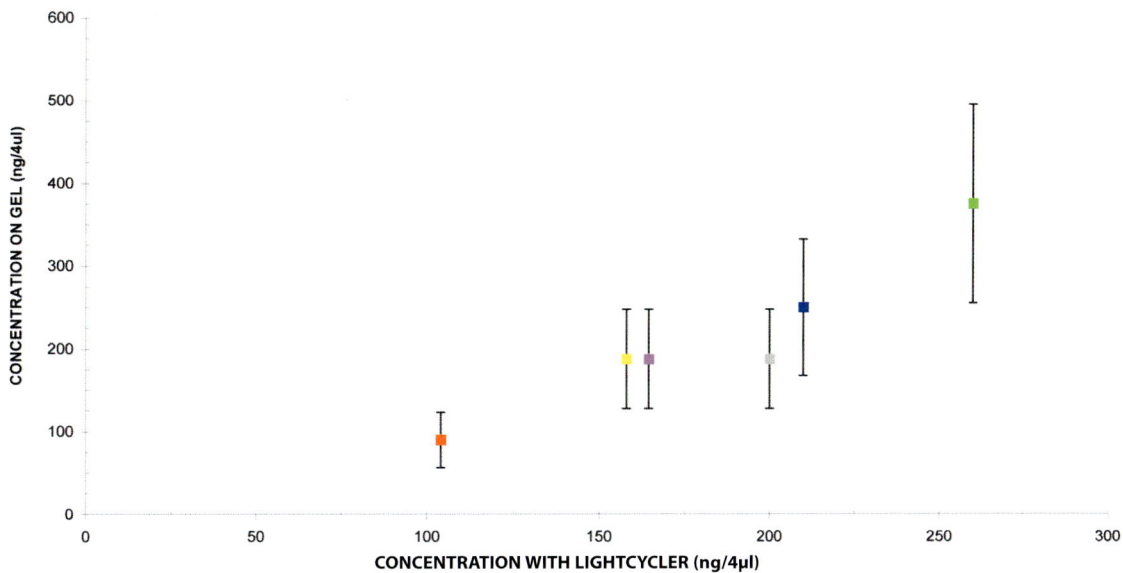

Fig. 5. Comparison of concentrations evaluated on gel and with lightcycler. Concentrations of six samples evaluated with LightCycler *(colored squares)* are plotted against concentrations evaluated on gel *(error bars)*

References

1. Sambrook J, Fritsch EF, Maniatis T (1989) Appendix E: commonly used techniques in molecular cloning. E5: quantitation of DNA and RNA. In: Molecular cloning. A laboratory manual, 2nd edn. Cold Spring Harbor Laboratory Press, Cold Spring Harbor
2. Fitzgibbon J, Gillett GT, Woodwark KJ, Boyle JM, Wolfe J, Povey S (1993) Mapping of *RXRB* to human chromosome *6p21.3*. Ann Hum Genet 57:203–209
3. Jabs WJ, Hennig H, Fittel M, Pethig K, Smets F, Bucsky P, Kirchner H, Wagner HJ (2001) Normalized quantification by real-time PCR of Epstein Barr virus load in patients at risk for transplant lymphoproliferative disorders. J Clin Microbiol 39:564–569

Genotyping by Guanosine-Dependent Quenching of Single-Labeled Fluorescein Probes

Andrew O. Crockett*

Introduction

Fluorescent-labeled oligonucleotide probes can be used in real-time PCR for rapid detection of specific product and for fine sequence analysis [1–5]. Changes in the magnitude of probe fluorescence upon hybridization to product DNA are responsible for the versatility of fluorescent oligonucleotides in monitoring PCR. Several techniques have been developed for using oligonucleotide probes to generate a fluorescent signal detectable by instruments such as the LightCycler. These fluorescent-labeled molecules utilize the natural phenomenon of fluorescence resonance energy transfer (FRET) to provide such a signal. During FRET, a donor fluorophore absorbs light energy and transfers it to either an acceptor, a second fluorophore that emits the light energy at a second, molecule-specific wavelength, or to a quencher, a molecule that dissipates the energy as heat. Exonuclease probes (TaqMan) [6], hairpin probes (Molecular Beacons) [7], and self-probing primers (Scorpions) [8] are examples of FRET systems in which the fluorescence of a donor is quenched by a second chromophore attached to the same molecule. Because two fluorophores are conjugated to a single oligonucleotide, the design, synthesis, and purification of these probes can be difficult and expensive.

Hybridization probe techniques eliminate many of the synthesis problems inherent to dual-labeled probes by utilizing two single-labeled oligonucleotides, one conjugated with the donor fluorophore and the other conjugated with an acceptor [1–5, 9]. These oligonucleotides are designed to anneal to adjacent regions of the PCR amplicon, bringing the attached fluorophores close enough together for FRET. Because each probe only requires conjugation with a single fluorophore, synthesis and purification of hybridization probes is simpler than for dual-labeled oligonucleotides. However, adjacent hybridization probes require a greater region of hybridization to the template DNA which makes the design prone to disruption by unexpected polymorphisms in the template. Also, signal generation is dependent upon two separate hybridization events, as both probes in a hybridization pair must anneal to a single template strand for the generation of the FRET signal.

---

* Andrew O. Crockett (✉) (e-mail: andrew.crockett@path.utah.edu)
  Department of Pathology, University of Utah School of Medicine, 50 North Medical Drive, Salt Lake City, UT 84132 USA

While developing a mutation detection assay on the LightCycler using hybridization probes, we noticed that one fluorescein-labeled probe produced a fluorescent signal strong enough to allow genotyping without addition of a second probe.

This genotyping probe produced a significant decrease in fluorescein emissions measured in F1 of the LightCycler upon annealing to complementary strand DNA. As the probe–template hybrid melted during temperature ramping, the fluorescein residue was released from quenching and the observed fluorescence emissions increased. Analysis of the fluorescence data throughout the melting transition produced melting troughs analogous to the melting peaks produced by hybridization probes upon melting (see Figs. 1–6). The observed fluorescein quenching effect was experimentally attributed to the interaction of the dye with complementary, native guanosine residues [10].

Fluorescein quenching by guanosine residues is an attractive technique for monitoring real-time PCR. Fluorescein quenching probes are inexpensive and simple to design, synthesize, and purify and are capable of fine sequence detection by melting curve analysis. In order to demonstrate the practicality of fluorescein quenching probes for genotyping, we developed assays to genotype:

1. The hemochromatosis mutation *C282Y* (*G845A*)
2. The hemochromatosis mutation *H63D* (*C187G*)
3. The cystic fibrosis-associated deletion *F508del*
4. The thermolabile mutation of methylenetetrahydrofolate reductase (*MTHFR; C677T*)
5. The factor V Leiden mutation (*G1691A*) responsible for heritable activated protein C resistance
6. The G20210A transition in the 3′-untranslated region of prothrombin

Materials

Equipment LightCycler Instrument (Roche Diagnostics, Mannheim, Germany)
LightCycler Capillary Tubes (Roche Diagnostics)

Reagents LightCycler-DNA Master Hybridization Probes (Roche Molecular Biochemicals, Mannheim, Germany)
Amplification primers (Operon Technologies, Alameda, CA, USA)
Fluorescein-labeled probes (Operon Technologies)

Sample Preparation

Procedure

Human genomic DNA was isolated from EDTA anticoagulated whole blood by phenol-chloroform extraction and ethanol precipitation [11]. After precipitation, the extracted DNA was resuspended in TE′, denatured by heating at 95°C for 5 min, and diluted to 1 OD (50 ng/μl). Extracted samples were frozen at –20°C or stored at 4°C until genotyping was performed.

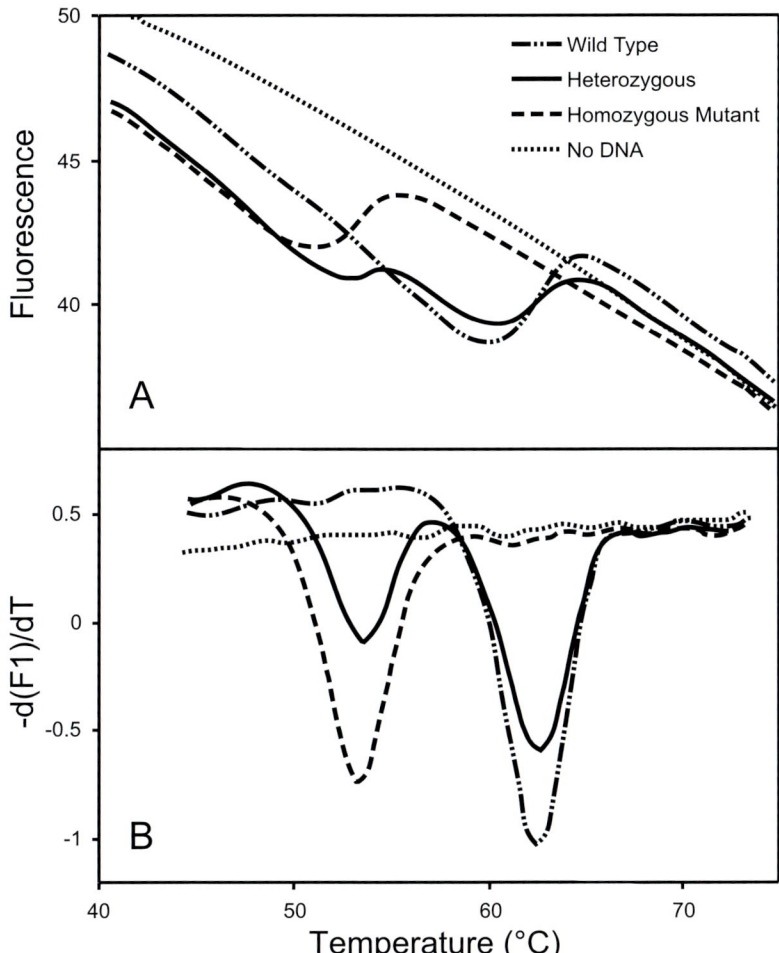

Fig. 1a,b. Homogeneous, real-time genotyping of hereditary hemochromatosis mutation *C282Y* (*G845A*) using a single fluorescein probe quenched by two guanosine residues. Melting curve data are presented as fluorescence vs temperature (**a**) and negative first-derivative vs temperature (**b**)

Oligonucleotides

Primers and probes were obtained from Operon Technologies and used without further purification. We synthesized 5′-fluorescein-labeled oligonucleotides with fluorescein-ON phosphoramidites (Clontech, Palo Alto, CA, USA) and blocked them from extension by a 3′-phosphate. Labeled probes were HPLC purified by the manufacturer. The extinction coefficient at 260 nm for each oligonucleotide was estimated from values for nucleotide pairs [12]. The concentration of unlabeled oligonucleotides was determined by absorbance at 260 nm. The concentration and purity of HPLC purified probes were assessed by measuring absorbance

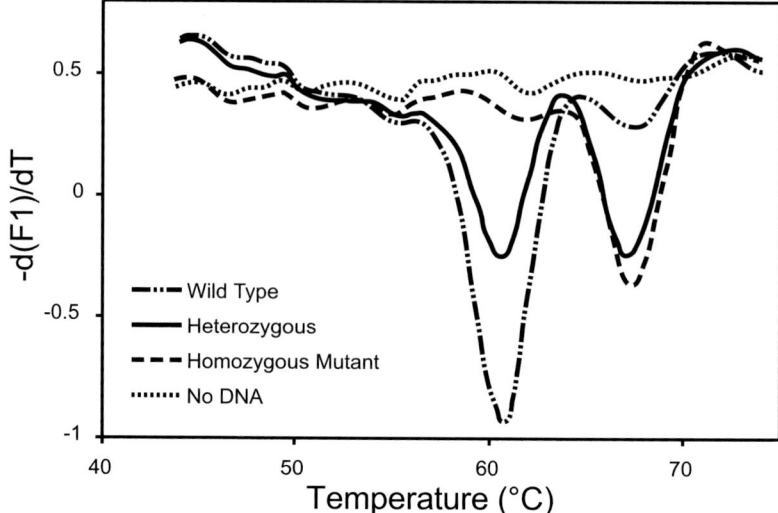

Fig. 2. Genotyping curves for hereditary hemochromatosis mutation *H63D* (*C187G*). The fluorescein probe was designed to be complementary to the mutant sequence. Quenching was provided by two complementary guanosines

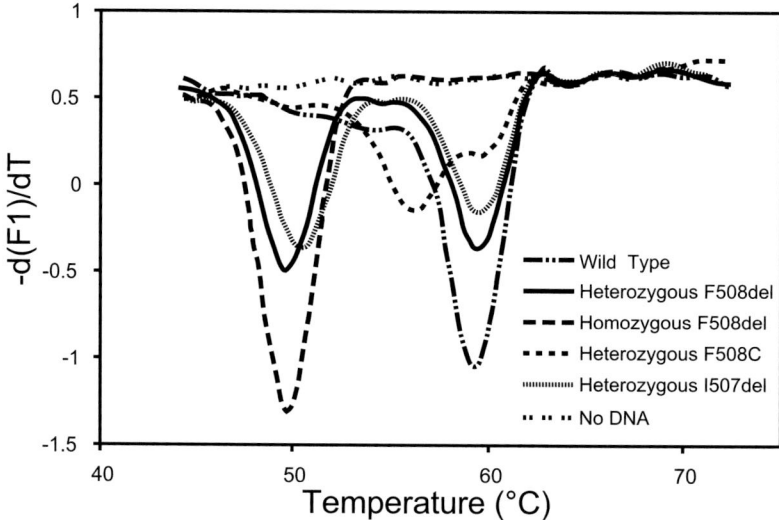

Fig. 3. Melting troughs for genotyping the cystic fibrosis-associated mutation *F508del*. Deletion of codon *508* results in a 3-bp mismatch loop in the probe, destabilizing the probe-template hybrid by 9°C. Quenching was provided by a single complementary guanosine

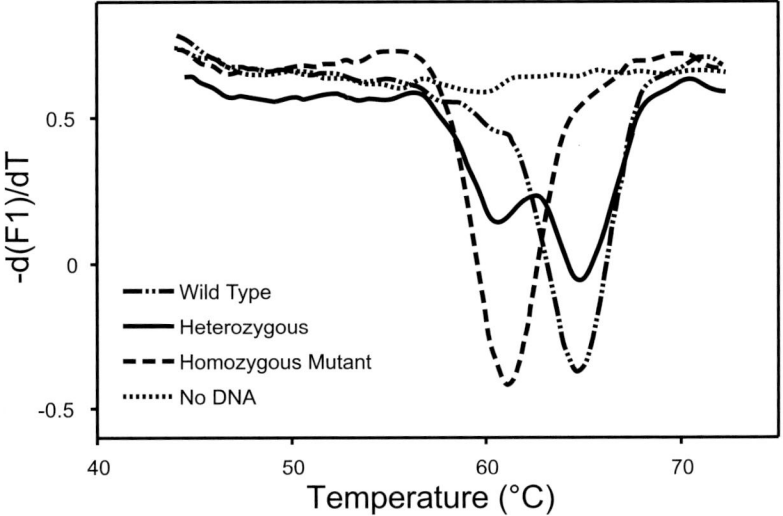

Fig. 4. Genotyping curves for the thermolabile mutation of *MTHFR* (*C677T*). The C:T transversion resulted in a G:T mismatch in the probe-template duplex. The G:T mismatch is one of the most stable known and destabilized the probe by only 3.9°C. Nevertheless, genotypes were clearly distinguishable. Quenching was provided by a single complementary guanosine

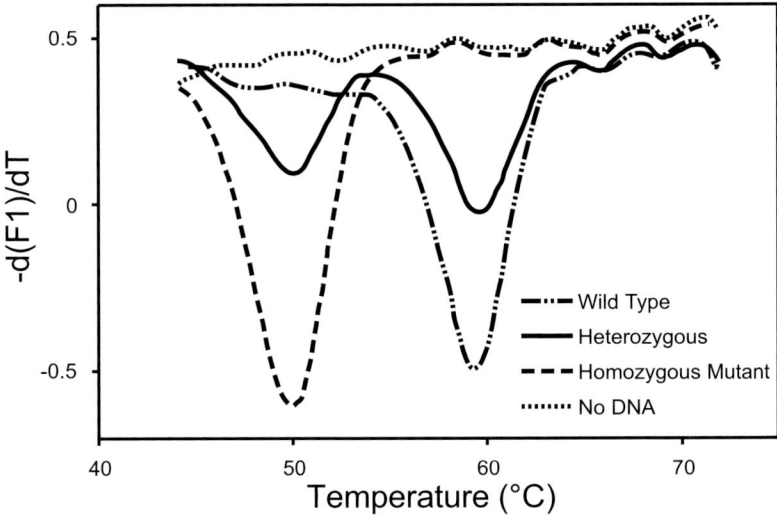

Fig. 5. Homogeneous genotyping of the factor V Leiden mutation by a single fluorescein probe. The fluorescein probe was designed to allow quenching by two complementary guanosine residues

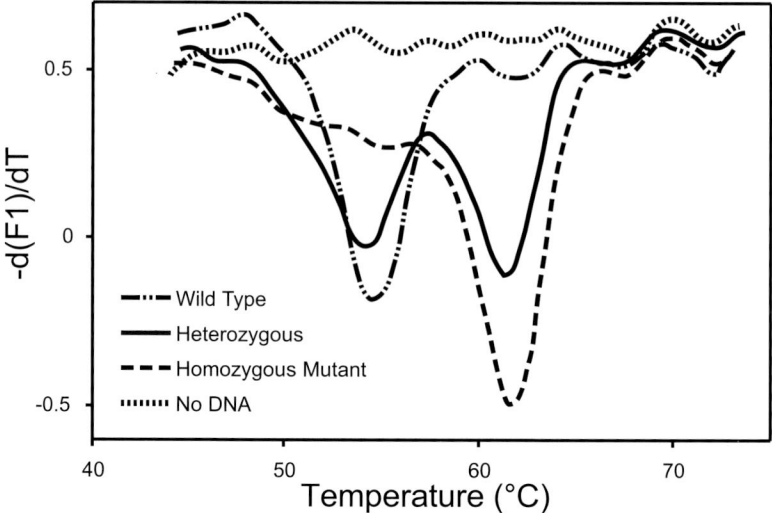

Fig. 6. Genotyping of the thrombosis-associated mutation *G20210A* in the 3′-untranslated region of prothrombin. The single fluorescein probe was designed to be complementary to the mutant sequence. A single complementary guanosine residue provided quenching

at 260 nm and 494 nm. The molar extinction coefficients for fluorescein at 260 nm and 494 nm were taken to be 12,000 M^{-1} cm^{-1} and 68,000 M^{-1} cm^{-1}, respectively [10]. Using these values, the ratio of the calculated concentrations of fluorescein to oligonucleotide was between 0.8 and 1.2 for each of the probes.

The primers used for product amplification were either taken from sequences previously described in the literature [1–4, 13, 14] or were newly designed for use with fluorescein quenching probes. The fluorescein-labeled genotyping probes were designed to maximize quenching by complementary guanosines. Because quenching relies upon the interaction between fluorophore and template, a critical factor in maximizing quenching is the position of the guanosine(s) relative to the fluorescent label. In order to quantify relative position versus quenching magnitude for complementary guanosines, a series of oligonucleotide template and probe sets were synthesized. Template sequences were designed to juxtapose single and multiple guanosine residues in varying positions opposite the fluorescein label. A single guanosine was able to quench a maximum of 25% of the fluorescein fluorescence when positioned as the first base lateral to the probe's region of hybridization on the template strand. Multiple guanosines increased quenching to approximately 40% [10]. With these design considerations in mind, probe sequences were devised which placed complementary guanosine residues in positions that provided strong quenching for each of the six loci. In each case, quenching was provided either by a single guanosine in the first lateral position or by two consecutive guanosines, with one positioned complementary to the last hybridizing base and the other as the first lateral base (Table 1).

Table 1. Oligonucleotides

Hereditary hemochromatosis C282Y (GenBank Accession #Z92910)				
	Position	Length	GC (%)	Tm (°C)
C282Y primers TGGCAAGGGTAAACAGATCC	6443	20	50.0	62.5
TACCTCCTCAGGCACTCCTC	6838 R	20	60.0	64.9
Product	6443–6838	396		
C282Y probe F-CACCTGGCACGTATATCTCTG-P	6729 R	21	52.4	62.8
Hereditary hemochromatosis H63D (GenBank Accession #Z92910)				
H63D primers CACATGGTTAAGGCCTGTTG	4620	20	50.0	62.2
GATCCCACCCTTTCAGACTC	4861 R	20	55.0	62.5
Product	4620–4861	242		
H63D probe F-CACACGGCGACTCTCATCATCATAGAAC-P	4779 R	28	50.0	68.4
Cystic fibrosis F508del (GenBank Accession #M55115)				
F508del primers GGAGGCAAGTGAATCCTGAG	245	20	55.0	63.2
CCTCTTCTAGTTGGCATGCT	500 R	20	50.0	62.6
Product	245–500	256		
F508del probe F-ATAGGAAACACCAAAGATGATATTTTC-P	451 R	27	29.6	60.6
Methylenetetrahydrofolate reductase C677T (GenBank Accession #U09806)				
C677T primers TGAAGGAGAAGGTGTCTGCGGGA	653	23	56.5	68.7
AGGACGGTGCGGTGAGAGTG	See [13]	20	65.0	68.7
Product		198		
C677T probe F-TGCGTGATGATGAAATCGGCTCC-P	695 R	23	52.2	66.5
Factor V Leiden (G1691A; GenBank Accession #L32764)				
G1691A primers TAATCTGTAAGAGCAGATCC	236	20	40.0	56.6
TGTTATCACACTGGTGCTAA	See [14][a]	20	40.0	58.8
Product		187		
G1691A probe F-CTGTATTCCTCGCCTGTC-P	276 R	18	55.6	59.9

Table 1. *Continued*

Prothrombin G20210A (GenBank Accession #M17262)				
	Position	Length	GC (%)	Tm (°C)
G20210A primers				
ATTGATCAGTTTGGAGAGTAGGGG	26667	24	45.8	63.9
GAGCTGCCCATGAATAGCACT	26820 R	21	52.4	65.8
Product	26667–26820	154		
G20210A probe				
F-TCTCAGCAAGCCTCAATGCT-P	26777	20	50.0	64.2

[a] Bertina et al. describe an intronic primer also found in PAC 86F14 (GenBank Accession #Z99572) position 62773. The paired exonic primer described here is located at position 62959 R of the PAC 86F14 sequence. The fluorescein probe is at position 62773.

LightCycler PCR

PCR conditions were optimized such that all six loci could be readily amplified and analyzed under the same protocol. One microliter (50 ng) of each DNA sample to be genotyped was added to 9 µl of master mix. The mixture was transferred to LightCycler capillary tubes, sealed, and centrifuged briefly to place the sample in the tip of the capillary. Tubes were loaded into a LightCycler carousel, transferred to the LightCycler instrument, amplified, and analyzed.

The following master mix was used:

	Volume [µl]	[Final]
LightCycler-DNA Master Hybridization probes	1	1×
MgCl$_2$ (25 mM)	0.8	3.0 mM
Forward primer (5 µM)	1	0.5 µM
Reverse primer (5 µM)	1	0.5 µM[a]
Fluorescein probe (1 µM)	1	0.1 µM[b]
dH$_2$O	4.2	
Total volume	9	

[a] The final concentration of reverse primer used in the assays for the cystic fibrosis-related mutation *F508del* and the hemochromatosis mutation *C282Y* was 0.06 µM.
[b] The final concentration of fluorescein probe used in detecting the *MTHFR C677T* mutation was 0.2 µM.

To allow uniformity between the six genotyping assays, a standard master mix was investigated. However, three assays required alteration of the primer or probe concentrations to allow optimal amplification and genotyping by fluorescein quenching. The assays designed to genotype the cystic fibrosis-related mutation *F508del* and the hemochromatosis mutation *C282Y* (*G845A*) were amplified using 0.5 µM forward primer and 0.06 µM reverse primer (8:1 asymmetry). This primer asymmetry proved to increase the magnitude of the fluorescence transitions during melting which, in turn, provided stronger melting peaks for genotyping. Another modification to the standard master mix was made in the assay designed to genotype the thermolabile mutation of *MTHFR* (*C677T*). This mutation was genotyped using twice the standard concentration of fluorescein-labeled probe (0.2 μM). The addition of extra probe allowed more robust identification of two separate melting transitions (i.e., melting peaks) in heterozygous samples (see Fig. 4).

- Amplification

Parameter	Value		
Cycles	50		
Type	Quantification		
	Segment 1	Segment 2	Segment 3
Target temperature [°C]	95	55	72
Incubation time [s]	0	10	0
Temperature transition rate [C/s]	20	20	1
Acquisition mode	None	Single	None
Gains	F1=20		

- Melting Curve Analysis

Parameter	Value		
Cycles	1		
Type	Melting curves		
	Segment 1	Segment 2	Segment 3
Target temperature [°C]	95	40	80
Incubation time [s]	0	30	0
Temperature transition rate [C/s]	20	20	0.1
Acquisition mode	None	None	Continuous
Gains	F1=20		

The following melting temperatures were found:

Locus	Allele	Pairing	T_m (observed)
HLA-H C282Y	WT	C-G match	62.5
	G845A	C-A mismatch	53.5
HLA-H H63D	WT	C-C mismatch	60.7
	C187G	C-G match	67.3
CFTR	WT	AAG-TTC match	59.4
	F508del	AAG-ΔΔΔ mismatch	49.6
	I507del	ATG-ΔΔΔ mismatch	50.4
	F508C	A-G mismatch	56.2
MTHFR	WT	G-C match	64.8
	C677T	G-T mismatch	60.9
Factor *V*	WT	C-G match	59.6
	G1691A	C-A mismatch	50.0
Prothrombin	WT	A-C mismatch	54.5
	G20210A	A-T match	61.5

Results

During the melting transition, fluorescein quenching probes produce a detectable increase in raw fluorescence as the probe melts from the template. This increase in probe fluorescence is converted into readily interpretable melting curves by the standard LightCycler Data Analysis software (LCDA). Because LCDA plots the negative first derivative of fluorescence vs temperature, the increase in raw fluorescence results in negative melting troughs, not positive peaks (see Figs. 1–6).

Fluorescein quenching by guanosine residues was used to genotype the *C282Y* hemochromatosis mutation as shown in Fig. 1. Fluorescence data collected in F1 of the LightCcycler during the melting protocol are plotted in Fig. 1a as fluorescence vs temperature. The plot shows an increase in fluorescence corresponding to the probe melting transition. A plot of the negative derivative of fluorescence vs temperature (–d(F1)/dt) for the melting curve is shown in Fig. 1b. The resulting melting troughs were readily interpretable for genotyping. Homozygous samples that were complementary to the probe (wild type) melted in a single transition at a relatively high temperature. Homozygotes that were mismatched to the probe sequence (homozygous mutant) melted in a single transition at a relatively low temperature. Heterozygotes (heterozygous) melted in two transitions. Quenching was provided by two consecutive complementary guanosines with one positioned as the last hybridizing base and the other as the first lateral base.

Other assays that use fluorescein quenching for genotyping are shown in Figs. 2–6. Melting troughs for genotyping the hemochromatosis mutation *H63D* are shown in Fig. 2. The fluorescein-labeled probe used to genotype the *H63D* locus was designed to be perfectly complementary to the mutant sequence (*C187G*) and made use of quenching by two complementary guanosines positioned as the first lateral base and the last hybridizing base.

Figure 3 shows melting troughs for genotyping the cystic fibrosis-associated mutation *F508del*. The three base pair deletions resulted in a mismatched 3-bp loop in the probe that destabilized the probe-template hybrid by approximately 10°C. Quenching was provided by a single complementary guanosine positioned as the first lateral base to the region of hybridization. While genotyping samples for the *F508del* mutation, two other variant alleles were identified whose mutations resulted in melting troughs distinguishable from *F508del*. One variation was the *F508C* mutation caused by a T:G transversion within codon *508*. This mutation resulted in an A:G mismatch that destabilized the probe-template hybrid by approximately 3°C. The second allele identified was a pathologic *I507del* variant. The deletion of codon *507* resulted in a 9°C shift from wild type and was differentiable from the *F508del* mutation by a 1°C difference in T_m. Both variant alleles were identified in heterozygotes.

Melting troughs from the genotyping assay for the thermolabile mutation of *MTHFR* (*C677T*) are shown in Fig. 4. The C:T transversion resulted in a G:T mismatch in the probe-template hybrid. The G:T mismatch is one of the most thermostable known and resulted in a T_m shift of only 3.9°C. A single complementary guanosine positioned as the first lateral base provided quenching for genotyping.

Results for the factor V Leiden genotyping assay are shown in Fig. 5. Quenching was provided by two consecutive guanosine residues positioned as the first lateral base and the last hybridizing base. Figure 6 shows melting troughs for the prothrombin G20210A genotyping assay. The probe was designed to be a perfect match for the mutant allele. A single complementary guanosine positioned as the first lateral base provided quenching.

Comments

Fluorescein quenching by complementary guanosine residues provides a robust means of genotyping PCR product by melting curve analysis. The key to obtaining an adequate signal for genotyping is the design of the fluorescein probe. The fluorescein label must be positioned such that it can readily interact with guanosines on the complementary strand. Empirical data from synthetic probe and template pairs suggest that the optimal position for a single complementary guanosine is as the first base lateral to the region of hybridization. In general, single guanosines positioned as the last hybridizing base or the second, third, or fourth lateral bases also contribute to quenching; however, their effect is weaker. Complementary guanosines internal to the last hybridizing base in the region of hybridization do not exhibit quenching [10].

When two or more guanosine residues are available for quenching, the interaction between the fluorophore and the complementary guanosines is greater and the observed signal quenching is stronger. For targets with two consecutive guanosines near the fluorescein label, the greatest quenching (approximately 32%) is demonstrated when the probe is designed to place the two residues as the first two lateral bases outside the region of hybridization. An alternate probe design for targets with two consecutive guanosines places the two residues as the first lateral base and the last hybridizing base with only a slight loss of quenching from the aforementioned design (approximately 29% quenching). This alternate design was used in genotyping assays for the factor V Leiden and hemochromatosis *H63D* and *C282Y* mutations described in this chapter. Additional complementary guanosine residues can increase the observed quenching to about 40% (targets with four consecutive lateral guanosines); however, this increase is not necessary for robust detection on sensitive instruments such as the LightCycler [10].

Guanosine residues at the end of the probe near the fluorescein label can contribute to permanent quenching of the fluorescent dye and should be avoided [15, 16].

Gain adjustments are important to collecting optimal fluorescence data for melting curve analysis. In most systems using fluorescein quenching probes, peak fluorescence is observed during the first few cycles of PCR. This is contrary to systems using hybridization probes in which fluorescence values increase during amplification. Because maximum fluorescence is seen early in PCR and decreases during cycling, gain adjustments may be necessary to keep fluorescence values from reaching low levels during amplification.

References

1. Bernard PS, Ajioka RS, Kushner JP, Wittwer CT (1998) Homogeneous multiplex genotyping of hemochromatosis mutations with fluorescent hybridization probes. Am J Pathol 153:1055–1061
2. Gundry CN, Bernard PS, Herrmann MG, Reed GH, Wittwer CT (1999) Rapid F508del and F508C assay using fluorescent hybridization probes. Genet Test 3:365–370
3. Bernard PS, Lay MJ, Wittwer CT (1998) Integrated amplification and detection of the C677T point mutation in the methylenetetrahydrofolate reductase gene by fluorescence resonance energy transfer and probe melting curves. Anal Biochem 255:101–107
4. Lay MJ, Wittwer CT (1997) Real-time fluorescence genotyping of factor V Leiden during rapid cycle PCR. Clin Chem 43:2262–2267
5. Von Ahsen N, Schutz E, Armstrong VW, Oellerich M (1999) Rapid detection of prothrombotic mutation of prothrombin (G20210A), factor V (G1691A), and methylenetetrahydrofolate reductase (C677T) by real-time fluorescence PCR with the LightCycler. Clin Chem 45:694–696
6. Lee LG, Connell CR, Bloch W (1993) Allelic discrimination by nick-translation PCR with fluorogenic probes. Nucl Acids Res 21:3761–3766
7. Tyagi S, Kramer FR (1996) Molecular beacons – probes that fluoresce upon hybridization. Nat Biotechnol 14:303–308
8. Whitcombe D, Theaker J, Guy SP, Brown T, Little S (1999) Detection of PCR products using self-probing amplicons and fluorescence. Nat Biotechnol 17:804–807
9. Wittwer CT, Herrmann MG, Moss AA, Rasmussen RP (1997) Continuous fluorescence monitoring of rapid cycle DNA amplification. Biotechniques 22:130–138
10. Crockett AO, Wittwer CT (2001) Fluorescein-labeled oligonucleotides for real-time PCR: using the inherent quenching of deoxyguanosine nucleotides. Anal Biochem 290:89–97
11. Thomas SM, Moreno RF, Tilzer LL (1989) DNA extraction with organic solvents in gel barrier tubes. Nucl Acids Res 17:5411
12. Borer PN (1975) In: Fasman, GD (ed) Handbook of Biochemistry and Molecular Biology, 3rd edn. (Nucleic acids, vol 1), CRC Press, Boca Raton, pp 589
13. Frosst P, Blom HJ, Milos R, Goyette P, Sheppard CA, Matthews RG, Boers GJ, den Heijer M, Kluijtmans LA, van den Heuvel LP, Rozen, R (1995) A candidate genetic risk factor for vascular disease: a common mutation in methylenetetrahydrofolate reductase. Nat Genet 10:111–113
14. Bertina RM, Koeleman BPC, Koster T, Rosendaal FR, Dirven RJ, deRonde H, van der Velden PA, Reitsma PH (1994) Mutation in blood coagulation factor V associated with resistance to activated protein C. Nature 369:64–67
15. Livak KJ, Flood SJA, Marmaro J, Giusti W, Deetz K (1995) Oligonucleotides with fluorescent dyes at opposite ends provide a quenched probe system useful for detecting PCR product and nucleic acid hybridization. PCR Meth Appl 4:357–362
16. Singh KK, Rucker T, Hanne A, Parwaresch R, Krupp G (2000) Fluorescence polarization for monitoring ribozyme reactions in real time. BioTechniques 29:344–351

Limitations of Melting Curve Analysis Using SYBR Green I – Fragment Differentiation and Mutation Detection in the CFTR-Gene

S. Kleinle*, K. Tabiti and S. Gallati

Introduction

Melting curve analysis with the LightCycler instrument can be performed with the SYBR Green I dye or with fluorescent hybridization probes.

Using the SYBR Green I dye, melting curve analysis offers the opportunity to identify and characterize PCR products with respect to their melting behavior. Each double-stranded (ds) DNA product has a specific melting temperature (T_m) at which 50% of the DNA is single-stranded. Continous monitoring of the denaturation process detects rapid loss of fluorescence near the melting point (T_m) and results in single, sharp melting peaks when plotted as the first negative derivative of fluorescence versus temperature (-dF/dT) [1]. As the T_m of a fragment is a function of the fragment length and the G+C content it can be used e.g. to distinguish amplification products from short fragments, such as primer dimers.

Using hybridization probes, melting curve analysis monitors the hybridization of the probes. Melting points of the hybridization probes are also dependent on probe length and G+C content. Moreover due to their small size, they can detect small mutations [2, 3]. A single base mismatch or small deletions within the probe-binding region will decrease the probe's melting point. Hence wildtype, mutant, and heterozygote samples can be distinguished by different melting points.

We wanted to examine the potential and limitations of fragment differentiation on melting curve analysis using SYBR Green I dye. We amplified various sequences of the cystic fibrosis transmembrane conductance regulator (CFTR)-gene [4] of increasing length and G+C content, respectively. Our results demonstrate a clear distinction of the different amplification products.

To test limitations of the SYBR Green I dye for the detection of small sequence differences, such as small mutations, we amplified fragments from the CFTR-gene comprising different mutations. We analyzed DNA samples from homozygote wildtype (wt) controls, homozygote mutant patients, and heterozygote patients. From homozygote wildtype and mutant samples homoduplices are formed on melting curve analysis while from heterozygote samples heteroduplices between wildtype and mutant DNA strands arise. We were able to detect 3bp-deletions,

---

* Stephanie Kleinle (✉) (e-mail: skleinle@gmx.de)
Laboratory Dr. Dr. Nevinny-Stickel, Josephspitalstr. 15, 80331 München, Germany

such as the common ΔF508- and the ΔI507-mutations, in fragments up to 90bp by heteroduplex formation. A distinction of the ΔF508- and ΔI507-homoduplices from wildtype-homoduplices was possible in fragments up to a length of 76bp.

In addition, we tested the 1bp-insertion 3905insT and the point mutation G542X (G→T) in small fragments (<80bp). We did not detect a significant difference for the 3905insT in a hetero- and a homozygote sample. For the G542X a heterozygote sample was analyzed and a small shift of T_m was detected.

In conclusion, we showed that the melting curve analysis of the LightCycler System using the SYBR Green I dye can be used for fragment differentiation and is even capable to detect small mutations in small fragments.

Materials

Equipment LightCycler Instrument (Roche Diagnostics, Switzerland)
Oligo primer analysis software 6 for Macintosh (MedProbe, Norway)

Reagents Oligonucleotides (Microsynth, Switzerland)
LightCycler-DNA Master SYBR Green I (Roche Diagnostics)
DNA extraction spin columns (Qiagen, Switzerland)

Procedure

Sample preparation Human total DNA was extracted from peripheral blood cells via DNA extraction spin columns (Qiagen, Switzerland). DNA was used at a concentration of 25 ng/μl.

Primer design Primers (Table 1) were designed and tested for primer dimers using the oligo primer analysis software (MedProbe, Norway). Amplification sequences with different GC content were designed to carry equal GC distributions. Primers for mutation detection were selected to carry the mutation in the middle of the amplification fragment.

Table 1. Oligonucleotides

CFTR Accession No.: M55107, M55115, M55116, M55127, M55129				
	CFTR exon	length	GC (%)	T_m (°C)
Primers AGTTTTCCTGGATTATGCCT CTTTGATGACGCTTCTGTAT		20 20	40 40	58.0 57.0
Product 1	10	90	36	76.8
Primers GAATTTCATTCTGTTCTCAGT CTTACCTCTTCTAGTTGGCA		21 20	33 45	55.2 57.7
Product 2	10	130	37	79

Table 1. *Continued*

CFTR Accession No.: M55107, M55115, M55116, M55127, M55129				
	CFTR exon	length	GC (%)	T_m (°C)
Primers				
GAGCCTTCAGAGGGTAAAAT		20	45	58.4
GGGTTCATATGCATAATCAAA		21	33	55.3
Product 3	10	200	37	80.1
Primers				
GTGAATATCTGTTCCTCCTC		20	45	56.1
TATGTCTGACAATTCCAGGC		20	45	58.8
Product 4	2	90	42	80.2
Primers				
ATGGAGACCAAATCAAGTGAA		21	38	58.6
TCAGCAGAATCAACAGAAGG		20	45	59.0
Product 5	2	135	42	80.9
Primers				
TTCAAATGGTGGCAGGTAGT		20	45	60.7
CCACAAGGACAAAGTCAAGC		20	50	60.8
Product 6	22	200	42	82
Primers				
TCTTCTCTAACTGCAGGTTG		20	45	58.3
CTGAAACTCACACTGGATCC		20	50	59.3
Product 7	22	200	49	86.3
Primers				
TTATGCCTGGCACCATTAAA		20	40	59.0
CGCTTCTGTATCTATATTCATC		22	36	55.1
Product 8	10	69	35	76
Primers				
TTATGCCTGGCACCATTAAA		20	40	59.0
TTGATGACGCTTCTGTATCT		20	40	57.8
Product 9	10	76	34	77.8
Primers				
GGATCAGGGAAGAGTACTT		19	47	56.6
GATTTCTCCTTCAGTGTTCA		20	40	56.4
Product 10	20	63	40	–
Primers				
GGATCAGGGAAGAGTACTT		19	47	56.6
CGATCTGGATTTCTCCTTCA		20	45	58.2
Product 11	20	70	41	–
Primers				
CTCCAAGTTTGCAGAGAAAG		20	45	58.1
CTCCACTCAGTGTGATTCC		19	53	58.7
Product 12	11	62	44	79.02

Melting temperatures (T_m) of the oligonucleotides were calculated using the nearest-neighbour method [7]. Melting temperatures of the products are mean experimental values.

LightCycler-PCR SYBR Green Master Mix for each 20 µl reaction:

	Volume µl	Final
LC-DNA Master SYBR Green I	2	1x
MgCl$_2$ (25 mM)	1.6	3 mM
Primers (10 µM each)	1+1	0.5 µM each
H$_2$O	12.4	
Total master mix volume per reaction	18	

18 µl of master mix and 2 µl DNA (25 ng/µl) were added to each capillary. Sealed capillaries were centrifuged and placed into the LightCycler rotor.

The following LightCycler-protocol was used for amplification of fragments ≥90bp. For amplification of the fragments of 69bp and 76bp incubation time was reduced to 5 s.

- Denaturation for 120 s at 95°C.
- Amplification

Parameter	Value		
Cycles	40		
Type	Quantification		
	Segment 1	Segment 2	Segment 3
Target temperature [°C]	95	55	72
Incubation Time (s)	0	5	8
Temperature Transition Rate (°C/s)	20	20	20
Acquisition Mode	None	Single	None

- Melting Curve Analysis

Parameter	Value		
Cycles	1		
Type	Melting curve		
	Segment 1	Segment 2	Segment 3
Target temperature [°C]	95	65	95
Incubation Time (s)	0	30	0
Temperature Transition Rate (°C/s)	20	20	0.1
Acquisition Mode	None	None	Cont.

Fluorimeter gain of channel 1 was set to 1.

Results

Fragment differentiation

We amplified fragments from the CFTR-gene of 90bp, 130bp, and 200bp length with different G+C contents (Table 1). For fragments of 36% and 37% GC, melting temperature (T_m) rose from 76.8°C to 79°C and to 80.1°C (Fig. 1A). Given the same G+C content, melting point temperatures rose with increasing length of the fragments. When length increased from 90bp to 135bp and 200bp for fragments of 42% GC, T_m rose from 80.2°C, to 80.9°C, and to 82°C with increasing length (Fig. 1B). For this amplifications no primer dimers were observed.

Melting behavior of sequences can be predicted by empirical formulas, which show a greater influence on T_m by the G+C content than by the fragment length (Fig. 2) 5. Therefore a larger temperature range can be covered by varying the G+C content than by changing the fragment length. Given the fragment length is more than doubled from 90bp to 200bp, T_m rose from 80.2°C to 82°C for a fragment of 42% GC. Whereas an increase of the G+C content from 37% to 42% and 49% for a 200bp fragment, T_m rose from 80.1°C to 82°C and 86.3°C (Fig. 1A and B).

Intra-assay and inter-assay variation

To confirm reproducibility, analyses were performed in replicates with two different DNA samples (for the ΔI507-, 3905insT, and G542X-mutations, respectively, only one sample was available) and amplifications were repeated three times. For each run melting curve analysis was repeated four times. We found an intra-assay variation for T_m of 0.08°C and an inter-assay variation of 0.5°C.

Mutation detection

We wanted to test sensitivity of melting curve analysis for the detection of small sequence differences. Therefore we amplified fragments of exon 10 of the CFTR-gene of 69bp, 76bp, and 90bp (products 8, 9, and 1; Table 1) carrying the 3bp-deletions, the ΔF508 and ΔI507-mutation. We analyzed DNA samples from homozygote wildtype (wt) controls, homozygote mutant patients, and heterozygote patients (Fig. 3A and B and 4). Mutations had been detected by single-strand-conformation-polymorphism (SSCP) analysis [6] and confirmed by direct sequencing.

Denaturation and renaturation of amplification products from samples heterozygous for a mutation results in double-stranded DNA: the homoduplices wt/wt and mutant/mutant, and the so called heteroduplices wt/mutant. Detection of a mutation by SYBR Green I depends on the differences of the melting temperatures of the different double-stranded DNA molecules.

Amplification of ΔF508- and ΔI507-heterozygotes resulted in heteroduplex formation which could be detected on melting curve analysis as an additional melting peak (Fig. 3A and B). These heteroduplices probably carry a loop with the surplus nucleotides, which decreases their T_m. Melting points of the ΔF508/wt- and ΔI507/wt-heteroduplices were about 3°C lower than melting points of wt/wt-homoduplices for the 69bp as well as for the 76bp amplification product. For amplifications of a 90bp-fragment this difference of T_m decreased to T_m=0.6°C. However, the ΔF508/wt-heteroduplex was still detectable by the LightCycler software (Fig. 4).

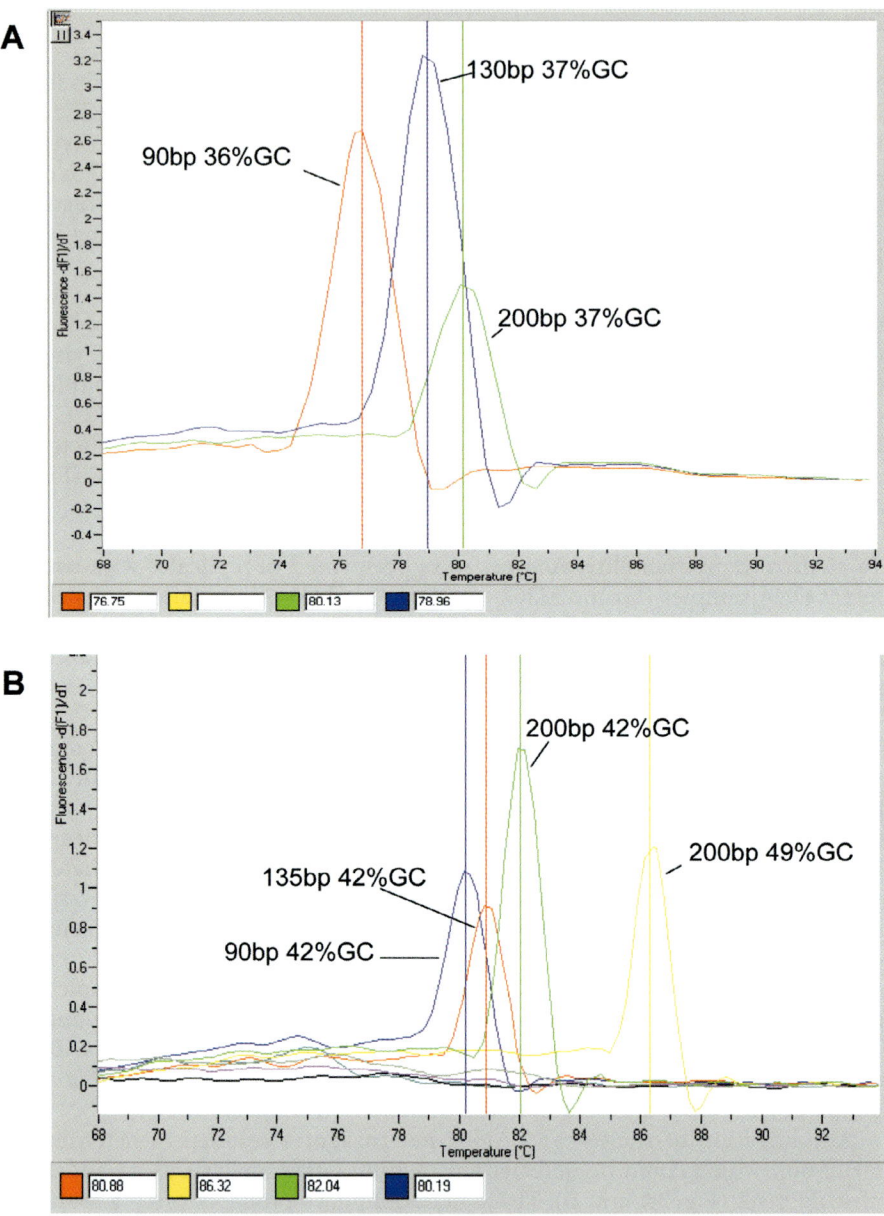

Fig. 1A, B. Melting curves of CFTR-fragments of different length and %GC. A: product 1 (red), product 2 (blue), product 3 (green); B: product 4 (blue), product 5 (red), product 6 (green), product 7 (yellow), H$_2$O-controls (pink, grey, anthracite, and black)

Limitations of Melting Curve Analysis Using SYBR Green I – Fragment Differentiation and Mutation Detection

$$T_m = 81.5 + 16.6 * \lg([salt] / (1 + 0.7 * [salt])) + 0.41 * (\%GC) - (500/L) + (2.09 * e^{-1.18 * SYBR\ dilution})$$

$$[salt] = [Na/K] + 4[Mg^{++} - dNTP]^{0.5} + [Tris^+],\ \text{for buffers at 3mM}\ Mg^{++}\ [salt] = 0.20$$

$$L = \text{fragment length}$$

Fig. 2. Relation between fragment length, GC content [5], and melting point (T_m)

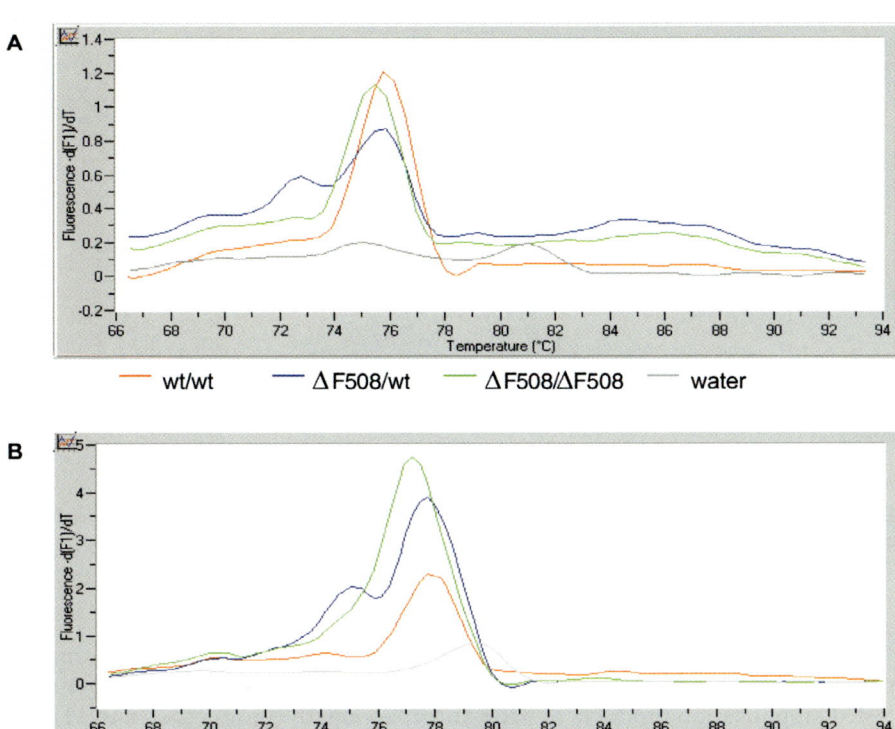

Fig. 3A, B. Detection of the 3bp-deletions ΔF508 (A) and ΔI507 (B) by melting curve analysis using SYBR Green I dye in the 69bp amplification fragment (product 8) (A) and the 76 bp amplification fragment (product 9) (B)

Amplification of the ΔF508- and ΔI507-homozygotes resulted in a slightly lower T_m compared to wildtype-homozygotes. The T_m-shift was about 0.5°C for the 69bp- as well as the 76bp-fragments (Fig. 3A and B). This difference was not detectable in the 90bp-fragment anymore.

Moreover, in fragments larger than 90bp (products 2 and 3) we did not detect any melting differences for wildtype samples as compared to ΔF508- and ΔI507-homo- or heterozygotes.

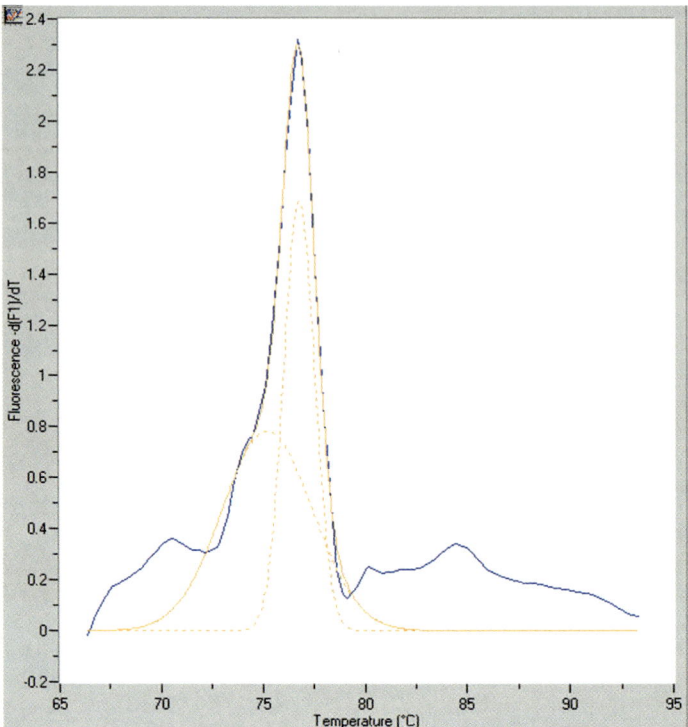

Fig. 4. Detection of the heteroduplex from ΔF508/wt heterozygotes in a 90bp fragment (product 1) on melting curve analysis. Blue line: fluorescence measured; brown dotted line: melting peaks as detected by the LightCycler analysis software

In addition, we analyzed a hetero- and a homozygote DNA sample from CF-patients who carried the 1bp-insertion 3905insT in exon 20 of the CFTR-gene. We tested amplification products of 63bp and 70bp length (products 10 and 11) and did not detect a significant difference on melting curve analysis.

We also tested the point mutation G542X (G→T) in exon 11 of the CFTR-gene in a 62bp amplification product (product 12). A DNA sample from a heterozygote patient was available. One melting peak was detected which showed a decrease of T_m of 0.4°C as compared to wildtype-homozygotes (Fig. 5). This melting peak was interpreted as an overlay of the melting peaks from the heteroduplices, homoduplex wt, and homoduplex mutant.

Comments

Fragment differentiation

We found fragments of different length and G+C content could be distinguished by melting curve analysis. Multiple specific products may be separated from each other due to different T_m in most cases, eliminating the need for gel electrophoresis.

Limitations of Melting Curve Analysis Using SYBR Green I – Fragment Differentiation and Mutation Detection

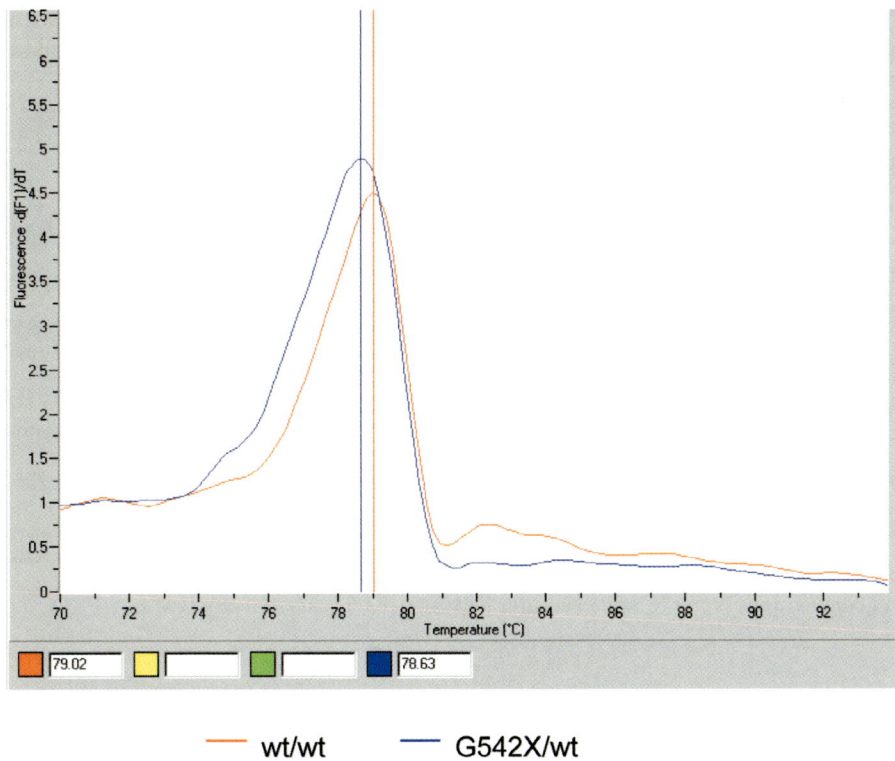

Fig. 5. Melting curve of the 62bp-fragment (product 12) carrying the point mutation G542X (G→T) amplified from a wildtype control and a patient heterozygote for the G542X

Amplification sequences were selected to contain equal G+C distributions. It has to be considered that G+C stretches in fragments may also lead to different melting curves.

Mutation detection

SYBR Green I dye may be used for mutation detection in small fragments (<80bp). We analyzed CF-mutations and were able to detect the 3bp-deletions ΔF508 and ΔI507 as well as the point mutation G542X (G→T). However, we did not find any difference for the 1bp-insertion 3905insT. 3bp-deletions were detectable in homo- as well as heterozygotes. The identification of point mutations seems to depend on the detection of the heteroduplices. Melting differences of heteroduplices compared to homoduplices are probably influenced by both the mutation and the surrounding sequence, e.g. for the 3905insT-mutation where one additional T is inserted into a stretch of 6 Ts. In addition, it has to be considered that mutation detection may also depend on the localization of the mutation within the amplification fragment. For all examined mutations here, the mutation site was located in the middle of the amplification fragment.

As the T_m-shift is mainly caused by the heteroduplices, mixing of samples with wildtype-DNA is recommended to create heteroduplices in homozygote mutant samples.

References

1. Ririe KM, Rasmussen RP, Wittwer CT (1997) Product differentiation by analysis of DNA melting curves during the polymerase chain reaction. Anal Biochem 245:154–160
2. Lay MJ, Wittwer CT (1997) Real-time fluorescence genotyping of factor V Leiden during rapid-cycle PCR. Clin Chem 43:2262–2267
3. Aoshima T, Sekido Y, Miyazaki T, Kajita M, Mimura S, Watanabe K, Shimokata K, Niwa T (2000) Rapid detection of deletion mutations in inherited metabolic diseases by melting curve analysis with LightCycler. Clin Chem 46:119–122
4. Zielenski J, Rozmahel R, Bozon D, Kerem B, Grzelczak Z, Riordan JR, Rommens J, Tsui LC (1991) Genomic DNA sequence of the cystic fibrosis transmembrane conductance regulator (CFTR) gene. Genomics 10:214–228
5. Wetmur JG (1995) in Molecular Biology and Biotechnology (Meyers, RA, Ed.), pp.605–608, VCH, New York
6. Liechti-Gallati S, Schneider V, Neeser D, Kraemer R (1999) Two buffer PAGE system-based SSCP/HD analysis: a general protocol for rapid and sensitive mutation screening in cystic fibrosis and any other human genetic disease. Eur J Hum Genet 7:590–598
7. Bommarito S, Peyret N, Santa Lucia J Jr (2000) Nucl Acids Res 28:1929–1934

SYBR Green I Analysis of the Trinucleotide Repeat Responsible for Huntington's Disease

CAMERON N. GUNDRY*, CARL T. WITTWER

Introduction

In 1993 the etiology of Huntington's disease (HD) was revealed as an expansion of the trinucleotide sequence CAG on chromosome 4p16.3. Normal individuals have between 10 and 29 repeats. An intermediate range has been established between 30 and 35 CAG repeats. Clinical manifestation of the disease occurs with 36–121 repeats, the median being about 44 [Huntington's Disease Collaborative Reasearch Group, 1993]. Since the causal discovery of the CAG repeat, PCR amplification and subsequent post-PCR gel sizing of the product has been the gold standard for the diagnosis of HD [Andrew, 1994]. We present a rapid PCR protocol wherein the T_m of the PCR product of the CAG repeat region can be used for Huntington's diagnosis. Melting curve analysis in the presence of SYBR Green I is used to determine the product T_m that is dependent on product length. In addition, we compared melting curves obtained on the ABI 5700 with those obtained on the LightCycler instrument.

Materials

LightCycler instrument, software and capillaries	**Equipment**
Oligonucleotide primers (Operon Technologies, Alameda, CA)	**Reagents**
Fluorescein-labeled primers (Operon Technologies, Alameda, CA)	
DMSO (Fisher Scientific Company, New Jersey, USA)	
LightCycler dNTPs (RMB)	
10 mM Magnesium Chloride buffer (Idaho Technology, UT, USA)	
KlenTaq™ polymerase (AB Peptides, St. Louis, MO, USA)	
Polymerase diluent (Idaho Technology, UT, USA)	
SYBR Green I (Molecular Probes, Eugene, OR)	

---

* Cameron N. Gundry (✉) (email: Cameron.gundry@path.utah.edu)
  Department of Pathology, University of Utah Medical School, Salt Lake City, UT 84132

Procedure

Sample Preparation

The DNA was extracted from peripheral blood leukocytes using standard phenol-chloroform DNA extraction and ethanol precipitation [Thomas, 1989] and stored in TE' (10mM Tris, pH 8.0, 0.1mM EDTA) at -20C. After extraction, the purified genomic DNA was boiled for 5 minutes in a water bath. DNA from 9 individuals with Huntington's disease and 18 normal individuals from prior studies [Bernard, 1999; Wittwer, 1993] was analyzed.

Primer Design

The amplification primers extend into the CAG repeat region several base pairs on both ends. They exclude the adjacent polymorphic CCG trinucleotide region as suggested previously [Andrew, 1994; Rubenstein, 1993].

Oligonucleotides

Table 1. Oligonucleotides

IT-15 (HD) Gene, Exon 1 (Genbank Accession #NM_002111)				
	Position	Length	GC (%)	T_m (°C)
Primers				
cgagtccctcaagtccttccagca	348	24	58.3	57.4
GTGGCGGCTGTTGCTGCT	442R	18	66.7	58.6
Product	348–442	56–179		

LightCycler PCR

The following master mix was used:

	Volume [µl]	[Final]
dNTP's (RMB, 2mM each)	1.00	0.20 mM each
MgCl$_2$ buffer (10 mM)	1.00	1.0 mM
DMSO (100%)	0.70	7.0 %
KlenTaq polymerase (diluted to 0.50U/ul)	1.0	0.05 units/l
Primers (5.0 M each)	1+1	0.50 M each
SYBR Green I (diluted 3000:1)	1	1:30,000
H$_2$0 (PCR grade)	2.3	–
Total volume	9	–

9 µl of master mix and 1µl of DNA (10–50ng/l) were added to each capillary. The capillaries were spun briefly and placed into the LightCycler carousel.

The following amplification protocol was used:

Parameter	Value	
Cycles	35	
Type	Quantification	
	Segment 1	Segment 2
Target temperature [°C]	97	68
Incubation time [s]	0	25
Temperature transition rate [°C/s]	20	20
Acquisition mode	None	Single
Display mode	F1	

The following melting peak protocol was used:

Parameter	Value		
Cycles	1		
Type	Melting Curves		
	Segment 1	Segment 2	Segment 3
Target temperature [°C]	95	60	95
Incubation time [s]	0	0	0
Temperature transition rate [°C/s]	20	0.20	0.05
Acquisition mode	None	None	Cont.
Display mode	F1		

To assign accurate lengths to products the samples were amplified according to the above protocols with the inclusion of a 5'-fluorescein-labeled primer used at 0.50 uM. The capillary tops were removed and each sample was spun down into a microfuge tube. The samples were sized on a commercial sequencing apparatus [Genomics Core Facility, University of Utah]. The lengths of the two PCR products in each sample were determined.

Results

A gradient of product T_ms was observed due to the polymorphic size of the CAG repeat region. The smallest normal allele observed was at 56bp ($CAG_n = 8$) while the largest normal allele observed was 113bp ($CAG_n = 27$). The corresponding T_ms were 79.82°C and 84.77°C, respectively. The smallest expanded (disease-causing) allele was 155 bp ($CAG_n = 40$) while the largest expanded allele was 179bp ($CAG_n = 48$). The corresponding T_ms were 85.85°C and 86.61°C. We observed a 1.08°C separation between the longest normal allele (27 CAG repeats) and the shortest Huntington disease allele (40 CAG repeats). All 9 HD samples had two widely separated peaks that indicated heterozygous composition. The T_ms of the 18 alleles from the 9 HD samples correlated with the respective lengths ($R^2=0.99$) as shown in Fig. 1.

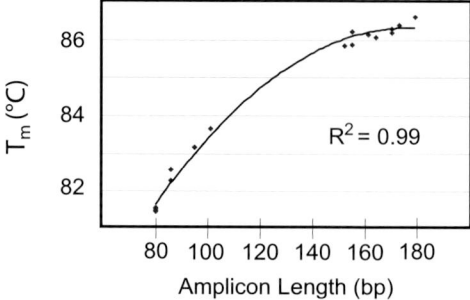

Fig. 1. T_m-length correlation in heterozygous Huntington's disease samples. T_ms of each allele in heterogygous Huntington's samples are plotted against amplicon length. Note the wide separation in T_m between the normal alleles (lower group) and the disease-causing ($CAG_n > 39$) alleles (upper group). The line represents a second degree polynomial fit to the data points. The T_m-length dependence is less pronounced as the product increases in size

In a within-run 10 tube replicate experiment we found a high degree of T_m reproducibility with a standard deviation of 0.053°C. The average T_m separation between a normal and an expanded allele in the heterozygous HD samples was 4.04°C, the median T_m separation was 4.32°C. The LightCycler software assigned 2 peaks, representing each allele, in all of the Huntington's samples and in 8 of 22 control samples. 5 of the 22 control samples were homozygous by product length analysis on the sequencing gel. Therefore, the heterozygote detection by T_m analysis was 8/17 (47%) in the control samples. These identified heterozygote samples were substantially different in size (mean = 18bp, median = 16.5bp). In the 9 heterozygous samples where the LightCycler instrument failed to identify 2 peaks, the difference in PCR product size was small (mean = 4.3bp, median = 4.5bp). When the 8 heterozygous normal samples that were assigned 2 peaks were included, the correlation between T_m and product length decreased only slightly (Fig. 2).

Comments

A T_m difference of 1 ° C was found between the longest normal sample (27 CAG repeats) and the shortest Huntington's disease sample (40 CAG repeats) in our data set. It would be rare to find a normal sample with more than 27 CAG repeats. Only 6 out of 600 control alleles from people without Huntington's disease were found to have >27 CAG repeats in a large population study (Kremer, 1994). Melting analysis can be used to screen for HD. We recommend that previously sized controls corresponding to 27, 30, and 35 CAG repeats should be included in the same run as the unknowns. Any sample with a T_m above the 27 repeat sample should be sized on a sequencing gel.

When the Huntington's disease samples were amplified in the LightCycler intrument and then melted on both the ABI 5700 heat block instrument and the LightCycler instrument, similar results were obtained (Fig. 3).

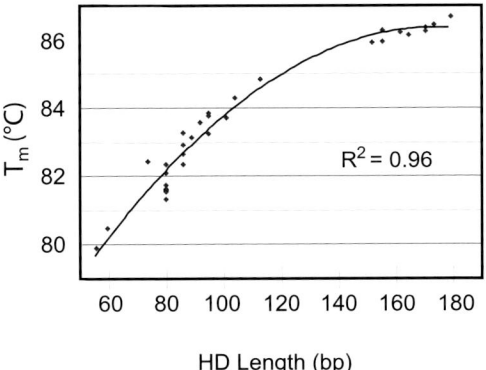

Fig. 2. T_m-length correlation in control and Huntington's disease samples. In addition to the Huntington's disease samples in Fig. 1, 16 more alleles (from 8 heterogygous normal samples) are included. Note that the lower end of the graph includes two unusually short PCR products. Also, an unsully long but non-disease-causing product is present (113 bp)

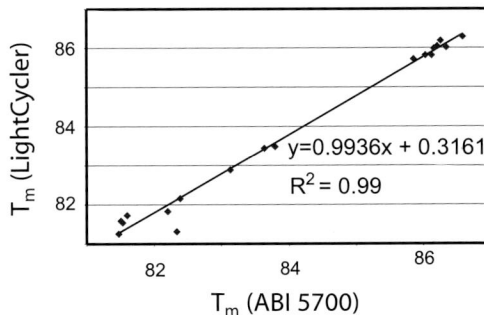

Fig. 3. Correlation of T_ms obtained on the LightCycler instrument and ABI 5700. Huntington's disease samples were first amplified on the LightCycler instrument and then melting curves obtained on both LightCycler and ABI 5700. T_ms were calculated from derivative melting curve data by non-linear least squares fitting of multiple Gaussian curves using custom software developed to accept data from both instruments

The LightCycler software is more flexible than the ABI 5700 software. This flexibility can be important in detecting heterozygotes. With SYBR Green I detection, as the cycle number increases, the higher temperature peak is favored over the lower temperature peak (Fig. 4). Therefore, the number of cycles before the melting curve is obtained during amplification is important. If the samples vary in starting template concentration or the efficiency varies between tubes, optimal melting results would be expected after different numbers of cycles. One solution to this problem is to acquire multiple melting curves during amplification, for example, at 30, 35, and 40 cycles.

This increases the ability of the LightCycler instrument to detect multiple peaks in a heterozygous sample. The melting protocols within such a run are dif-

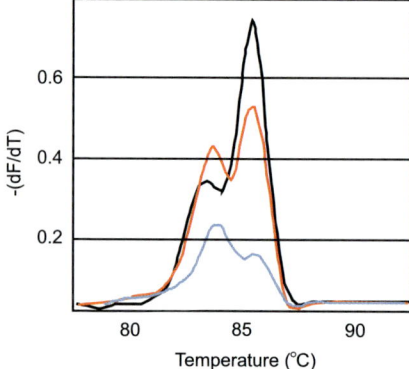

Fig. 4. Effect of amplification cycle number on derivative melting curves of PCR products using SYBR Green I. The relative area of the two peaks changes as PCR amplification progresses. At 35 cycles (blue) the lower peak is predominant while both lower and upper peaks are of approximate equal size at 40 cycles (red). At 50 cycles the upper peak (black) is much larger than the lower

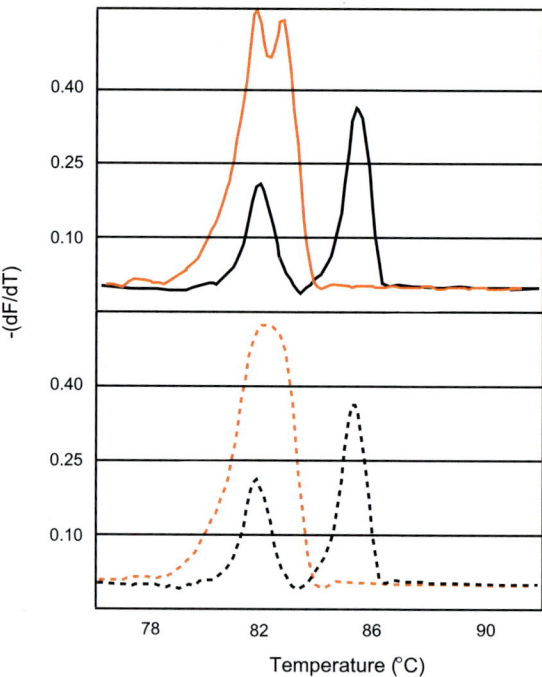

Fig. 5. Effect of the temperature interval analyzed on the resolution of derivative melting curves. Two closely-spaced peaks in a normal heterozygote are apparent when 1.1° C are averaged on the LightCycler software (solid red line). Only one peak is apparent when 3.1°C are averaged (dotted red line). In a Huntington's heterozygote, the alleles are widley separated in T_m and distinct peaks are apparent when analyzed either with 1.1°C (solid black line) or 3.1°C (dotted black line) intervals

ferent from the protocol in table 4 in that the target temperature in segment 2 is increased to 72°C to prevent primer extension during each melting curve.

In contrast, the control software of the ABI 5700 has a fixed melting protocol. With LightCycler, the user can also alter the temperature interval averaged when calculating derivative melting curves. If the interval is too large, separable peaks converge and sensitivity is lost (Fig. 5).

References

A novel gene containing a trinucleotide repeat that is expanded and unstable on Huntington's disease chromosomes. (1993). The Huntington's Disease Collaborative Research Group. Cell. 72(6):971–83.

Andrew SE, Goldberg YP, Theilmann J, Zeisler J, Hayden MR. (1994). A CCG repeat polymorphism adjacent to the CAG repeat in the Huntington disease gene: implications for diagnostic accuracy and predictive testing. Hum Mol Genet. 3(1):65–7.

Bernard PS, Pritham GH, Wittwer CT. (1999). Color multiplexing hybridization probes using the apolipoprotein E locus as a model system for genotyping. Anal Biochem. 273(2):221–8.

Kremer B, Goldberg P, Andrew SE, Theilmann J, Telenius H, Zeisler J, Squitieri F, Lin B, Bassett A, Almqvist E, et al. (1994). A worldwide study of the Huntington's disease mutation. The sensitivity and specificity of measuring CAG repeats. N Engl J Med. 330(20):1401–6.

Rubinsztein DC, Leggo J, Barton DE, Ferguson-Smith MA. (1993). Site of (CCG) polymorphism in the HD gene. Nat Genet. 5(3):214–5.

Thomas SM, Moreno RF, Tilzer LL. (1989). DNA extraction with organic solvents in gel barrier tubes. Nucleic Acids Res. 17(13):5411.

Wittwer CT, Marshall BC, Reed GH, Cherry JL. (1993). Rapid cycle allele-specific amplification: studies with the cystic fibrosis delta F508 locus. Clin Chem. 39(5):804–9.

Applications in Genetics

II

Parallel Genotyping of Different Genes: A Rapid Real-Time PCR Approach 67
STEFAN FRONHOFFS, THOMAS BRÜNING, HANS VETTER, YON KO

**Detection of a Single Base Substitution in Single Cells
by Melting Peak Analysis Using Dual-Color Hybridization Probes** 77
GERARD PALS

**Rapid Screening for Five Major Cystic Fibrosis Mutations
by Melting Peak Analysis Using Fluorogenic Hybridization Probes** 85
SIEGFRIED BURGGRAF, NAEEM MALIK,
EDITH SCHUHMACHER, BERNHARD OLGEMÖLLER

**LightCycler PCR for the Polymorphisms −308 and −238 in the *TNF Alpha* Gene
and for the TNFB1/B2 Polymorphism in the *LT Alpha* Gene** 95
LUKAS BESTMANN, NÄDER HELMY, FELICIA GAROFALO,
AYNUR DEMIRTAS, DIETER VONDERSCHMITT, FRIEDRICH E. MALY

**Rapid Genotyping of 2-bp and 9-bp Deletion Mutations
Using the LightCycler Instrument** 107
TSUTOMU AOSHIMA, MITSUHARU KAJITA, YOSHITAKA SEKIDO,
SHUNJI MIMURA, KAZUYOSHI WATANABE, KAORU SHIMOKATA,
TOSHIMITSU NIWA

**Genotyping of the Methionine-Valine Polymorphism at Codon 129
of the Human Prion Protein by Melting Point Analysis of Fluorescently Labeled
Hybridization Probes** 115
SIEGFRIED BURGGRAF, SIEGFRIED KÖSEL, SABINE LOHMANN,
REINHARD BECK, BERNHARD OLGEMÖLLER

**Rapid Detection of Missense Mutations in the Prostatic Steroid
5α-Reductase Gene Using Real-Time Fluorescence PCR
and Melting Curve Analysis** 129
MARKUS NAUCK, WINFRIED MÄRZ, HEINRICH WIELAND

Parallel Genotyping of Different Genes: A Rapid Real-Time PCR Approach

STEFAN FRONHOFFS*, THOMAS BRÜNING**, HANS VETTER*, YON KO*

Introduction

Much progress has been made in identifying clinically relevant point mutations in genomic DNA. As a result, procedures that allow fast, accurate and easy analysis of known point mutations are needed for diagnosis. To substantially increase the throughput of sample analysis and/or to analyze individual gene mutation patterns, it would be helpful to establish the analysis of different parameters in a parallel procedure.

Hereditary hemochromatosis (HH) is a common autosomal recessive disorder of iron metabolism, characterized by excessive iron deposition in a variety of organs leading to multiorgan dysfunction. Two point mutations, G845A and C187G, referred to as C282Y and H63D, have been detected in the recently identified hemochromatosis gene (*HFE*) [1]. Most patients (>88%) with HH were found to be homozygous for the C282Y mutation and a small number are heterozygous for both the C282Y and the H63D mutations [1, 2]. An α_1-antitrypsin (AAT) deficiency is an inherited cause of emphysema and cirrhotic liver disease [3]. The most common AAT-deficiency variant, PiZ, is characterized by a single G→A point mutation on exon V, resulting in an amino acid change at position 342 [4]. Factor V Leiden, a point mutation (G1691A) in the *factor V* gene, has been demonstrated to be one of the most frequent, inherited prothrombotic risk factors [5]. Familial defective apolipoprotein (apo) B100 is a genetic disorder in which a reduced affinity of low-density lipoprotein (LDL) to the LDL receptor results in hypercholesterolemia and premature atherosclerosis [6]. The disorder is caused by a G→A mutation in the codon for amino acid 3500 [7, 8]. Apolipoprotein E gene (*APOE*) polymorphism is characterized by three major allelic variants *APOE2*, *APOE3* and *APOE4*. *APOE3* is considered to be the wild-type allele. *APOE2* is characterized by the point mutation C4070T, which causes an amino

---

*   Stefan Fronhoffs, Hans Vetter, Yon Ko
    (✉) (e-mail: yonko@uni-bonn.de)
    Medizinische Universitäts-Poliklinik, Wilhelmstraße 35–37, 53111 Bonn, Germany

** Thomas Brüning
    Berufsgenossenschaftliches Forschungsinstitut für Arbeitsmedizin
    an der Ruhr Universität Bochum, Bürkle-de-la-Camp Platz 1, 44789 Bochum, Germany

acid change at position 158. *APOE4* alleles differ by the point mutation C3932T, causing an amino acid change at position 112 [9, 10]. The homozygous *APOE2* genotype leads to familial dysbetalipoproteinemia with increased plasma levels of cholesterol and triglycerides [11]. The *APOE4* genotype has been associated with decreased longevity, hypercholesterolemia and an increased prevalence of Alzheimer's disease [12].

To evaluate patients for the presence of these point mutations, we established a LightCycler-assisted polymerase chain reaction (PCR), which greatly simplifies the identification of all mutations in a parallel analysis within 60 min.

Materials

Equipment MagNa Pure LC, version 2.0 (Roche Diagnostics, Mannheim, Germany)
LightCycler instrument (Roche Diagnostics)
LightCycler software, vers. 3.5 (Roche Diagnostics)
LightCycler Carousel centrifuge (Roche Diagnostics)

Reagents MagNa Pure LC DNA Isolation Kit I (Roche Diagnostics)
LightCycler Factor V Leiden Mutation Detection Kit (Roche Diagnostics)
LightCycler Apo B Mutation Detection Kit (codon 3500) (Roche Diagnostics)
LightCycler Apo E Mutation Detection Kit (codons 112 and 158) (Roche Diagnostics)
AAT PCR primers and hybridization probes (TIB MOLBIOL, Berlin, Germany)
HFE (H63D and C282Y mutations) PCR primers (MWG-BIOTECH, Ebersberg, Germany)
HFE (H63D and C282Y mutations) hybridization probes (TIB MOLBIOL)

Procedure

DNA Extraction Genomic DNA was isolated from 200 µl peripheral blood anticoagulated with EDTA on the MagNa Pure LC instrument using the MagNa Pure LC DNA Isolation Kit I, according to the manufacturer's instructions. The isolated DNA was eluted in 200 µl elution buffer and a 5-µl aliquot was transferred to PCR.

Primers and Hybridization Probes For the detection of the *HFE* H63D mutation, PCR primers amplifying a 169-bp fragment of the *HFE* gene and two hybridization probes labeled with fluorescein and LightCycler Red 640 (LCRed640) were constructed as described in Table 1. The LCRed640-labeled probe hybridizes to the *HFE* sequence, which harbors the C187G mutation. For analysis of the *HFE* C282Y mutation, different PCR primers and two hybridization probes also labeled with fluorescein and LCRed640 were used as described in Table 1. The primer pair amplifies a 347-bp fragment of the *HFE* gene. The fluorescein-labeled hybridization probe hybridizes to the *HFE* sequence, which contains the G845A mutation. To identify the PiZ mutation, we used a primer pair to amplify a 182-bp fragment of exon V of the *AAT* gene (Table 1). The two hybridization probes were labeled with fluorescein and

Table 1. Oligonucleotides

α_1-Antitrysin gene (GenBank Accession # K02212)			
	Position	GC (%)	T_m (°C)
Primers			
GTGTCCACGTGAGCCTTGCTC	11840–11860	63.2	58.7
GTTTGTTGAACTTGACCTCGG	12001–12021	47.6	56.3
Probes			
CTTCAGTCCCTTTCTCGTCGATGGTC-F	11930–11955	53.8	64.9
LCRed640-CACAGCCTTATGCACGGCCTGGAG-P	11904–11927	62.5	69.5
Hemochromatosis gene, H63D mutation (GenBank Accession # Z92910)			
Primers			
GCCTCAGAGCAGGACCTTGG	4684–4703	65.0	63.5
CAGCTGTTTCCTTCAAGATGC	4973–4993	47.6	57.9
Probes			
CTTGAAATTCTACTGGAAACCCATGGAGTT-CGGGGCTCC-F	4779–4817	51.3	76.7
LCRed640-CACGGCGACTCTCATCATCATAG-AACACGAACA-P	4745–4777	48.5	70.4
Hemochromatosis gene, C262Y mutation (GenBank Accession # Z92910)			
Primers			
TGGCAAGGGTAAACAGATCC	6443–6462	50.0	57.3
CTCAGGCACTCCTCTCAACC	6813–6832	60.0	61.4
Probes			
AGATATACGTACCAGGTGGAG-F	6712–6732	47.6	50.2
LCRed640-CCCAGGCCTGGATCAGCCCCT-CATTGTGATCTGGG-P	6735–6769	62.9	80.5

LCRed640 and hybridized to the wild-type sequence with the fluorescein-labeled probe spanning the mutation site (Table 1). For detection of the factor V Leiden mutation, specific PCR primers and hybridization probes provided with the LightCycler Factor V Leiden Mutation Detection Kit (Roche Diagnostics) were used. The PCR primers amplify a 222-bp fragment of the *factor V* gene. The two hybridization probes included a fluorescein-labeled probe spanning the mutation site and hybridizing to wild-type and an adjacent LCRed640-labeled probe. For detection of the *APOB100* C9774T and G9775A mutations, primers and hybridization probes provided with the LightCycler Apo B Mutation Detection Kit (codon 3500) were used. The specific primers amplify a 207-bp fragment of the *APOB100* gene. The LCRed640-labeled hybridization probe hybridizes to the wild-type sequence, which harbors the mutation sites. For detection of the *APOE* C3932T and C4070T mutations on codon 112 and 158, respectively, primers and hybridization probes provided with the LightCycler Apo E Mutation Detection Kit (codon 112 and codon 158) were used. The specific primer pair amplifies a 265-bp fragment of the *APOE* gene. Two different hybridization probe pairs labeled with fluorescein and LCRed640 or fluorescein and LCRed705 were used to distinguish between the C3932T and C4070T mutations, respectively.

LightCycler Master Mix

To simultaneously screen genomic DNA samples for the described mutations, six different PCR reaction mixtures had to be prepared as follows. A negative control containing no genomic DNA was run with the samples. A positive control for each mutation analysis was performed as indicated in Fig. 1 with either heterozygous or heterozygous and homozygous DNA replacing the sample DNA. LightCycler-assisted real-time PCR of the *HFE* C282Y and H63D and the *AAT* gene mutations was performed in a final reaction volume of 20 µl, consisting of the following components:

	Volume [µl]	[Final]
LightCycler DNA Master Hybridization Mix	2	1×
MgCl$_2$ stock solution (25 mM)	2.4	4 mM
Primers (20 µM each)	1 + 1	1 µM each
Hybridization probes (3 µM each)	1 + 1	0.15 µM each
H$_2$O (PCR-grade)	6.6	
Genomic DNA	5	

LightCycler-assisted real-time PCR of the *factor V* and *APOB100* genes was performed in a final reaction volume of 20 µl consisting of the following components:

	Volume [µl]	[Final]
LightCycler Mutation Detection Mix	2	1×
LightCycler Reaction Mix	2	1×
H$_2$O (PCR-grade)	11	
Genomic DNA	5	

LightCycler-assisted real-time PCR of the *APOE* gene was performed in a final reaction volume of 20 µl consisting of the following components:

	Volume [µl]	[Final]
LightCycler Mutation Detection Mix	4	1×
LightCycler Reaction Mix	2	1×
H$_2$O (PCR-grade)	9	
Genomic DNA	5	

parallel real-time PCR

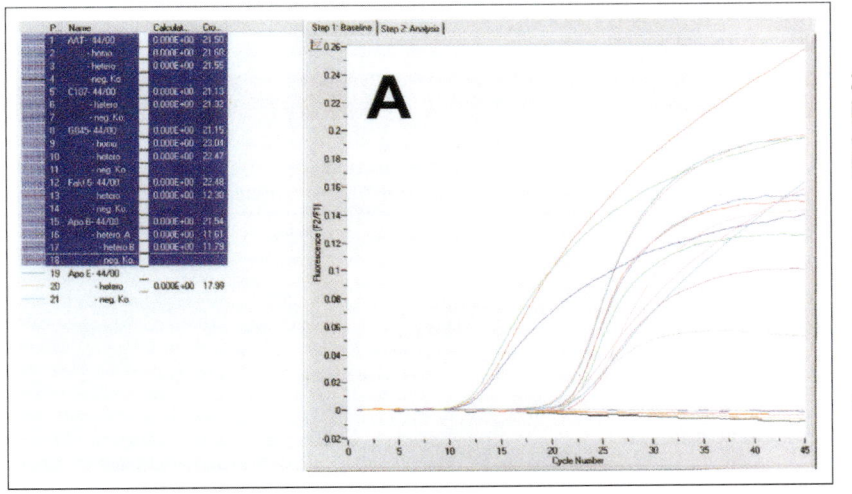

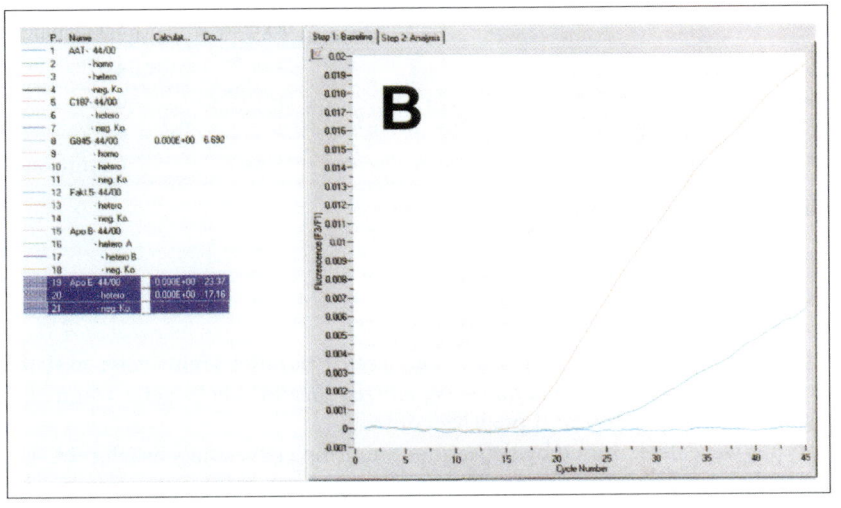

Fig. 1A, B. Online fluorescence curves of parallel LightCycler-assisted PCR amplification of *HFE*, *AAT*, *factor V*, *APOB100* and *APOE* gene fragments from a representative patient sample. Using LCRed640-labeled hybridization probes, fluorescence was measured in channel 2 (**A**). For the detection of the *APOE* C4070T mutation, LCRed705-labeled hybridization probes were used and fluorescence was measured in channel 3 (**B**). A negative control was run with each mutation analysis. As indicated in Fig. 2, positive controls for each mutation analysis were amplified as follows: heterogeneous control DNA in case of the *HFE* H63D, *factor V*, *APOB100* and *APOE* mutations, heterogeneous and homogeneous control DNA in case of the *HFE* C282Y and *AAT* mutations

Parallel LightCycler PCR Analysis

For the parallel LightCycler PCR mutation analysis, the following protocol was used:
- Denaturation at 95°C for 150 s
- Amplification

Parameter	Value		
Cycles	45		
Type	Quantification		
	Segment 1	Segment 2	Segment 3
Target temperature [°C]	95	55	72
Incubation time [s]	0	20	20
Temperature transition rate [°C/s]	20	20	20
Acquisition mode	None	Single	None
Gains	F1=1	F2=15	F3=30

- Melting curve analysis

Parameter	Value		
Cycles	1		
Type	Melting curves		
	Segment 1	Segment 2	Segment 3
Target temperature [°C]	95	45	85
Incubation time [s]	0	60	0
Temperature transition rate [°C/s]	20	20	0.1
Acquisition mode	None	None	Continuous

- Cooling at 40°C for 300 s

Results

Figure 1 displays the online fluorescence curves of parallel LightCycler-assisted PCR amplification of *HFE*, *AAT*, *factor V*, *APOB100* and *APOE* gene fragments from a sample with negative and positive controls.

Genotypes were determined by subsequent melting curve analysis (Fig. 2A–G), which revealed single-characteristic melting temperatures (T_m) associated with the wild-type and mutant alleles. As shown in Fig. 2A–F, the following T_m profile was found measuring the LCRed640 fluorescence in channel 2 (F2): the *HFE* H63D mutant and wild-type alleles were associated with a T_m of 73°C and 67.5°C, respectively (Fig. 2A); the *HFE* C282Y mutation and the wild-type alleles were associated with a T_m of 61.5°C and 56.0°C, respectively (Fig. 2B); in the case of the *AAT* gene, the PiM and the PiZ alleles were associated with a T_m of 69°C and 64°C, respectively (Fig. 2C); the mutated *factor V* allele showed a T_m of 57°C, whereas the wild-type allele was associated with a T_m of 65°C (Fig. 2D); the *APOB100* C9774T and G9775A mutant alleles and the wild-type allele were associated with

melting curves

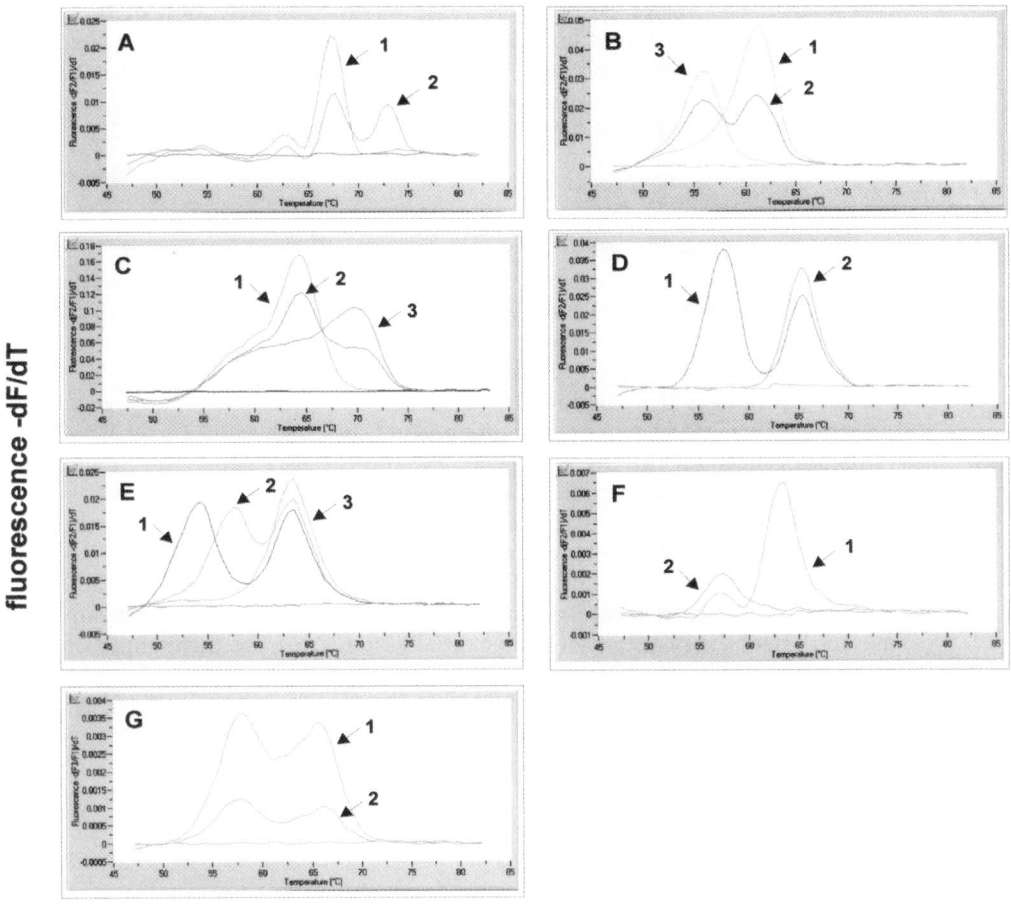

Fig. 2A–G. Melting curve analysis of the amplified gene fragments shown in Fig. 1. Melting curve analysis was performed by plotting the negative first derivative of the fluorescence (*F*) with respect to temperature (*T*) against temperature [-(dF/dT) vs T]. **A** Heterozygous *HFE* H63D mutant (*curve 2*, control DNA), homozygous *HFE* wild type (*curve 1*, sample DNA). **B** Homozygous *HFE* C282Y mutant (*curve 1*, control DNA), heterozygous *HFE* C282Y mutant (*curve 2*, control DNA), homozygous *HFE* wild type (*curve 3*, sample DNA). **C** Homozygous *AAT* PiZZ genotype (*curve 1*, control DNA), heterozygous *AAT* PiMZ genotype (*curve 2*, control DNA), homozygous *AAT* (PiMM) wild type (*curve 3*, sample DNA). **D** Heterozygous *factor V* mutant (*curve 1*, control DNA), homozygous *factor V* wild type (*curve 2*, sample DNA). **E** Heterozygous *APOB100* G9775A mutant (*curve 1*, control DNA), heterozygous *APOB100* C9774T mutant (*curve 2*, control DNA), homozygous *APOB100* wild type (*curve 3*, sample DNA). **F** Heterozygous *APOE* C3932T mutant (*curve 1*, control DNA), homozygous *APOE* C3932T mutant (*curve 2*, sample DNA). **G** Heterozygous *APOE* C4070T mutant (*curve 1*, control DNA), heterozygous *APOE* C4070T mutant (*curve 2*, sample DNA)

a T_m of 57°C, 53°C and 62.5°C, respectively (Fig. 2E); the *APOE* genotyping revealed T_ms of 56°C and 62.5°C for the C3932T mutant (APOE*E4) and the wild-type (APOE*E3) alleles, respectively (Fig. 2F). For the detection of the *APOE* C4070 T (APOE2*E2) mutation, LCR 705 fluorescence was measured in channel 3 (F3). Two T_ms of 57.5°C and 66°C were identified, corresponding to the mutant and the wild-type alleles, respectively (Fig. 2G). The patient's DNA was found to contain no mutation except for the heterogenous APOE*E2/E3 genotype.

Comments

With the introduction of real-time PCR technology in combination with melting curve analysis, a rapid and convenient method for the detection of single-nucleotide polymorphism became available. High throughput of sample analysis and no requirement of post-PCR sample processing combined with easy handling are key advantages. This methodical approach has already been applied for the detection of several known point mutations in genes such as the *factor V* and *prothrombin* genes [13, 14], the *HFE* gene [15] and the *AAT* gene [16]. In addition, commercially available mutation detection kits have been developed.

We describe here a LightCycler-assisted real-time PCR, which allows the simple, rapid and accurate evaluation of patients for the presence of the *HFE* C282Y, *HFE* H63D, factor V Leiden, *AAT* PiZ, *APOB100* C9774T and G9775A and *APOE* C3932T and C4070T mutations in a parallel analysis within 60 min. This improved method offers the possibility of substantially increasing the throughput of sample analysis, of analyzing different samples for the presence of any of these gene mutations and/or of determining the individual gene mutation pattern. In terms of the rapidity, flexibility and the subsequent economic advantages, this approach is suitable for use in clinical and routine laboratory applications.

References

1. Feder JN, Gnirke A, Thomas W, Tsuchihashi Z, Ruddy DA, Basava A, Dormishian F, Domingo R Jr, Ellis MC, Fullan A, Hinton LM, Jones NL, Kimmel BE, Kronmal GS, Lauer P, Lee VK, Loeb DB, Mapa FA, McClelland E, Meyer NC, Mintier GA, Moeller N, Moore T, Morikang E, Wolff RK et al (1996) A novel MHC class-I-like gene is mutated in patients with hereditary haemochromatosis. Nat Genet 13:399–408
2. Jouanolle AM, Fergelot P, Gandon G, Yaouanq J, Le Gall JY, David V (1997) A candidate gene for hemochromatosis: frequency of the C282Y and H63D mutations. Hum Genet 100:544–547
3. Crystal RG (1990) Alpha 1-antitrypsin deficiency, emphysema, and liver disease Genetic basis and strategies for therapy. J Clin Invest 85:1343–1352
4. Fabbretti G, Sergi C, Consales G, Faa G, Brisigotti M, Romeo G, Callea F (1992) Genetic variants of alpha-1-antitrypsin (AAT). Liver 12:296–301
5. Bertina RM, Koeleman BP, Koster T, Rosendaal FR, Dirven RJ, de Ronde H, van der Velden PA, Reitsma PH (1994) Mutation in blood coagulation factor V associated with resistance to activated protein C. Nature 369:64–67
6. Innerarity TL, Weisgraber KH, Arnold KS, Mahley RW, Krauss RM, Vega GL, Grundy SM (1987) Familial defective apolipoprotein B-100: low density lipoproteins with abnormal receptor binding. Proc Natl Acad Sci U S A 84:6919–6923

7. Soria LF, Ludwig EH, Clarke HR, Vega GL, Grundy SM, McCarthy BJ (1989) Association between a specific apolipoprotein B mutation and familial defective apolipoprotein B-100. Proc Natl Acad Sci U S A 86:587–591
8. Boren J, Ekstrom U, Agren B, Nilsson-Ehle P, Innerarity TL (2001) The molecular mechanism for the genetic disorder familial defective apolipoprotein B100. J Biol Chem 276:9214–918
9. Weisgraber KH, Rall SC Jr, Mahley RW (1981) Human E apoprotein heterogeneity. Cysteine-arginine interchanges in the amino acid sequence of the apo-E isoforms. J Biol Chem 256:9077–9083
10. De Knijff P, van den Maagdenberg AM, Frants RR, Havekes LM (1994) Genetic heterogeneity of apolipoprotein E and its influence on plasma lipid and lipoprotein levels. Hum Mutat 4:178–194
11. Smit M, de Knijff P, van der Kooij-Meijs E, Groenendijk C, van den Maagdenberg AM, Gevers Leuven JA, Stalenhoef AF, Stuyt PM, Frants RR, Havekes LM (1990) Genetic heterogeneity in familial dysbetalipoproteinemia. The E2(lys146–gln) variant results in a dominant mode of inheritance. J Lipid Res 31:45–53
12. Smith JD (2000) Apolipoprotein E4: an allele associated with many diseases. Ann Med 32:118–127
13. Lay MJ, Wittwer CT (1997) Real-time fluorescence genotyping of factor V Leiden during rapid-cycle PCR. Clin Chem 43:2262–2267
14. Van den Bergh FA, van Oeveren-Dybicz AM, Bon MA (2000) Rapid single-tube genotyping of the factor V Leiden and prothrombin mutations by real-time PCR using dual-color detection. Clin Chem 46:1191–1195
15. Parks SB, Popovich BW, Press RD (2001) Real-time polymerase chain reaction with fluorescent hybridization probes for the detection of prevalent mutations causing common thrombophilic and iron overload phenotypes. Am J Clin Pathol 115:439–447
16. Ortiz-Pallardo ME, Zhou H, Fischer HP, Neuhaus T, Sachinidis A, Vetter H, Bruning T, Ko Y (2000) Rapid analysis of alpha1-antitrypsin PiZ genotype by a real-time PCR approach. J Mol Med 78:212–216

Detection of a Single Base Substitution in Single Cells by Melting Peak Analysis Using Dual-Color Hybridization Probes

Gerard Pals*

Introduction

Detection of point mutations in very small amounts of DNA in our lab is performed in micro dissected material from tumors and small premalignant lesions to assess loss of mutant or wild-type alleles of tumor suppressor genes. Even more demanding is the detection of point mutations in a single cell, which is required in preimplantation genetic diagnosis.

Real-time PCR followed by melting curve analysis, using hybridization probes, is highly sensitive, rapid and an efficient approach to mutation detection. We have used this approach on the LightCycler instrument for the detection of single-base mutations in a single cell, without nested PCR [1]. The presence of mutant and wild-type alleles was assessed using a single FITC-labeled anchor probe and wild-type and mutant-specific detection probes with different labels (LCRed640 and LCRed705).

Hybridization probes were designed on sequences in the *BRCA1* gene, covering mutations that are present in cell lines available in our lab. PCR reactions of small fragments (100–300 bp) containing the probe sequences were optimized using the SYBR Green I, before using hybridization probes. The 5′-probes were 3′-labeled with FITC, whereas the 3′-probes, covering the mutation, were 5′-labeled with LCRed640 (wild-type probes) or LCRed705 (mutant probes). Dual-color detection of wild-type and mutant sequence in a single tube was tested on single cells. The reaction mix was prepared in reaction capillaries and a single lymphoblast cell, picked by manual micromanipulation, was added to this mix. The DNA from the cell is released during the 5-min preheating step of the PCR, using the FastStart hybridization kit (Roche Diagnostics, Mannheim, Germany). Reproducible results were obtained, without the need of nested PCR. The technique is useful for microdissected tumors and, with other genes, has great potential for preimplantation diagnosis in IVF and analysis of residual disease in cancer.

---

* Gerard Pals (✉) (e-mail: g.pals@vumc.nl), Department of Clinical Genetics, Vrije Universiteit Medical Center, Amsterdam, The Netherlands, Van der Boechorststraat 7, 1081 BT Amsterdam, The Netherlands

Materials

Equipment LightCycler instrument

Reagents QIAmp blood kit (Qiagen, Hilden, Germany)
DNAzol (Life Technologies, Bethesda, MD, USA)
PCR primers (Life Technologies)
Hybridization probes (TIB Molbiol, Berlin, Germany)
FastStart LightCycler kits (Roche Diagnostics)

Procedure

DNA Samples DNA, cell lines and paraffin-embedded tissue were anonymized samples from our diagnostic laboratory. Lymphoblast cell lines were produced from peripheral blood from patients and healthy blood donors, by EBV transformation. Tissue samples from patients were from tumor and normal breast tissue.

DNA was isolated form peripheral blood using DNAzol (Life Technologies, Bethesda, MD, USA) and from paraffin-embedded tissue using the QIAgen blood kit (Qiagen) according to the manufacturer's instructions.

PCR Primers and Hybridization Probes Primers and probes were designed using OLIGO5 software (MBI, Cascade, CO, USA). Some primers and probes are different from the ones we described before [1, 4]. The probes were designed to have melting temperatures (T_m) above the T_m of the PCR primers and the T_m of the fluorescein-labeled probe (anchor probe) above the T_m of the other probes. The probes were custom-made by TIB-MOLBIOL (Berlin, Germany). The sequences are given in Table 1.

Optimizing PCR PCR products for PCR optimization were prepared with DNA from peripheral blood lymphocytes from healthy control subjects (wild type) and from patients who were heterozygous for the mutations studied ($2841G{\rightarrow}T$ and $3159G{\rightarrow}T$ in *BRCA1*).

PCR reactions were performed according to the LightCycler kit instructions (FastStart DNA Master SYBR Green I), using master mixes supplied in kits from Roche Diagnostics with 5 picomoles of each primer per reaction in a 20-µl reaction volume (25µM). Using 10 ng of DNA per capillary, 35 PCR cycles were monitored with SYBR Green I. At the end of the PCR cycles, melting curve analysis (LightCycler kit instructions) was performed by heating to 95°C for 2 min, followed by cooling to 65°C and gradual heating to 92°C at 0.2°C/s. The PCR reactions were optimized with respect to annealing temperature and [Mg^{2+}], using SYBR Green I, final Mg^{2+} concentrations from 2–5 mM and annealing temperatures around the T_A suggested by the OLIGO5 software.

For hybridization probe experiments, PCR reactions were set up according to the LightCycler kit instructions (FastStart DNA Master Hybridization). Probes were used at final concentrations of 0.1–0.4 µM. PCR conditions were derived from the SYBR Green I experiments. Mg^{2+} was optimized again with the probes.

Table 1. Oligonucleotides

Oligonucleotides (5'-3') for *2841G→T* mutation in *BRCA1*				
	Position	Length	GC (%)	T$_m$ (°C)
GCT CCG TTT TCA AAT CCA G	2727	19	47.4	66.7
CAGTTTCGTTGCCTCCTCTGAAC	2990	20	50.0	67.0
Product		283	40.6	82.2
Hybridization probes				
AGAAACAAAGTCCAAAAGTCACTTTTGAAT-F	2797	30	30.0	72.6
Wild type: LCRed640-GAACAAAAGGAAGAA-AATCAAGGAA-p	2829	25	32.0	69.3
Mutant: LCRed705-GAACAAAAGGAATAAA-ATCAAGGAA-p	2829	25	21.8	67.3
Oligonucleotides (5'-3') for *3519G→T* mutation in *BRCA1*				
GGCCAAAATTGAATGCTATG	3349	20	40.0	66.5
TTGCAGTCAAGTCTTCCAAT	4025	20	40.0	66.4
Product		696	39.8	83.0
Hybridization probes (sense strand):				
Anchor: CTGTTAATACAGATTTCTCTCCA-TATCTGAT-F	3475	31	32.3	68.9
Wild type: LCRed640-TCAGATAACTTAGA-ACAGCCTATGG-p	3507	25	40.0	66.4
Mutant: LCRed705-TCAGATAACTTATAAC-AGCCTATGG-p	3507	25	36.0	64.3

PCR reaction mix for *3519G→T* mutation in *BRCA1*:

	Volume [µl]	[Final]
FastStart DNA Master Hybridization Mix	2	1×
MgCl$_2$ (20 mM)	2.4	4 mM
BRCA1 primers (5 µM)	1+1	0.25 µM
Anchor probe (2 µM)	1	0.1 µM
Wild-type probe (2 µM)	0.75	75 nM
Mutant probe (2 µM)	0.75	75 nM
Template (1 cell)	2	
HPLC grade water	9.1	

Applications in Genetics

The following PCR program was used for both tests:
- Denaturation at 95°C for 8 min
- Amplification

Parameter	Value		
Cycles	60		
Type	Quantification		
	Segment 1	Segment 2	Segment 3
Target temperature [°C]	95	57	72
Incubation time [s]	5	10	20
Temperature transition rate [°C/s]	20	20	20
Secondary target temperature [°C]	0	0	0
Step size	0	0	0
Step delay	0	0	0
Acquisition mode	None	Single	None
Gains	F1=1	F2=20	F3=40

- Melting Curve Analysis

Parameter	Value		
Cycles	1		
Type	Melting Curve		
	Segment 1	Segment 2	Segment 3
Target temperature [°C]	95	45	80
Incubation time [s]	10	20	0
Temperature transition rate [°C/s]	20	20	0.1
Secondary target temperature [°C]	0	0	0
Step size	0	0	0
Step delay	0	0	0
Acquisition mode	None	None	Continuous

- Cooling for 30 s at 40°C

Single-Cell Experiments Lymphoblast cell cultures were diluted in culture media (HAM-F10) on a glass microscope slide until separate cells could be easily distinguished. Single cells were selected under visual inspection on an inverted microscope, using a microcapillary inspection pipette. Each cell was transferred to a reaction capillary containing the complete reaction mixture that had been prepared in advance and kept on ice.

PCR reaction mix for *2841* mutation:

	Volume [μl]	[Final]
FastStart DNA Master Hybridization mix	2	1×
MgCl$_2$ (20 mM)	2.4	4 mM
BRCA1 primers (5 μM)	1+1	0.25 μM
Anchor probe (2 μM)	1.5	0.15 μM
Wild-type probe (2 μM)	1	0.1 μM
Mutant probe (2 μM)	1	0.1 μM
Template (1 cell)	~2	
HPLC grade water	6.1	

The final volume was brought to 18 μl per capillary with nuclease-free water. The final reaction volume was around 20 μl, but varied depending on the volume of culture media transferred with the cell. The DNA from the cell becomes available for PCR after the preheat cycle of 8 min at 95°C, which is also needed to activate the hot-start polymerase.

Results and Discussion

PCR and mutation detection were optimized for two different mutations in the *BRCA1* gene: *2841GT* (recurrent in the Dutch population [2]) and *3519G→T* (found in several populations [3]; numbers according to cDNA sequence).

For the *2841GT* mutation in *BRCA1*, the homozygous wild-type PCR product showed a single peak at 59.2°C, whereas the heterozygous products (mutant) showed an additional peak at 54°C (Fig. 1a). The mutant-specific probe showed a mutant melting peak at 58.3°C and a wild-type peak at 52.4°C (Fig. 1b).

Comparable results were obtained with the *BRCA1 exon 11* PCR fragment containing the *3519G→T* mutation. In this case the homozygous wild type had a single peak at 64.1°C with the wild-type probe, whereas the heterozygous mutant/WT showed an additional peak at 60.2°C (Fig. 1c). The mutant probe showed a single peak at 57.2°C for the homozygous wild type and an additional peak at 63.7°C for the heterozygous sample (Fig. 1d). Both wild type probes show a shoulder in the wildtype sample. This may be due to misincorporations during PCR.

Mutation Detection in DNA from Blood

A single-base substitution was reproducibly detected in single cells from a cell line containing the *BRCA1 2841GT* mutation (Fig. 2). All cells in four series (three mutant cells per series) showed a heterozygous pattern with the wild type as well as mutant probe. In none of the experiments loss of an allele due to unequal PCR amplification was observed. The wild-type cells all showed a single wild-type peak with both probes. With the wild-type probe, the peak shows a shoulder towards lower temperature. This probably reflects PCR mutations during the early cycles of the reaction. The shoulder did not result in misinterpretation, but it

Mutation Detection in Single Cells

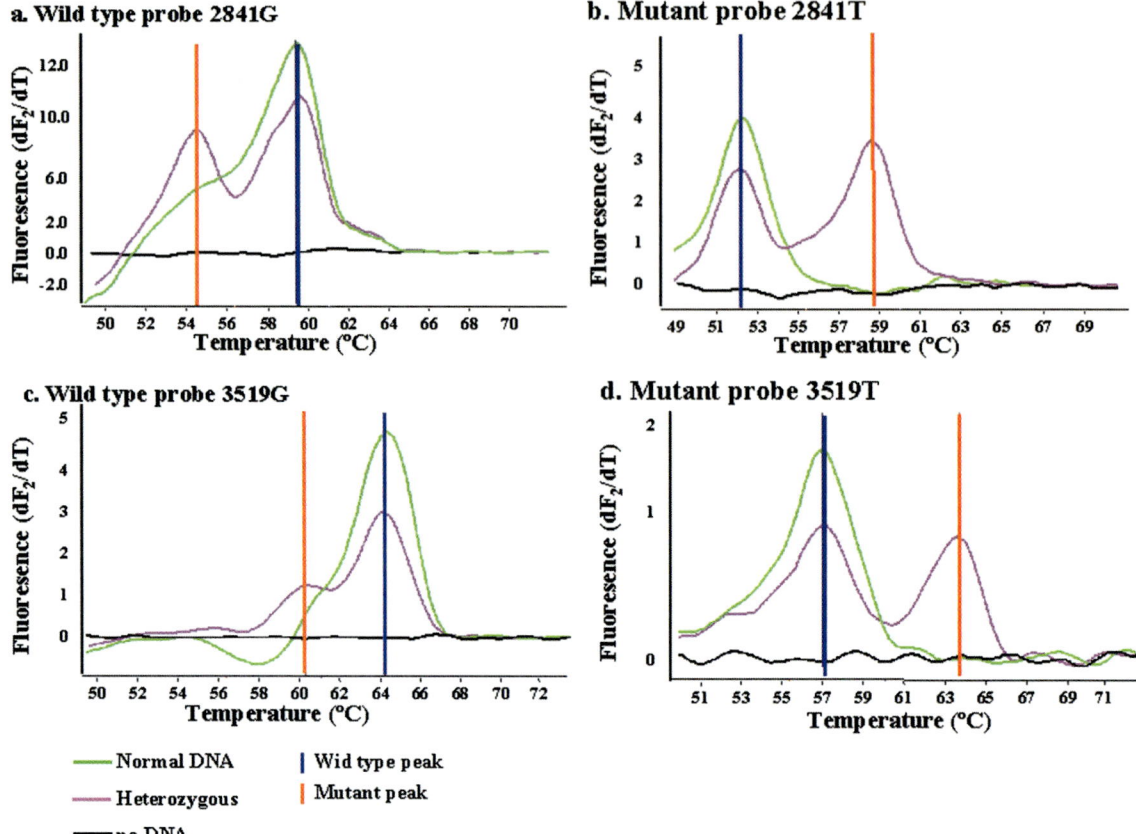

Fig 1. Detection of the 2841GT (fig. 1A, B) and 3591GT (fig. 1 C, D) mutations in DNA from blood from a normal control person and from two carriers of a mutation. The heterozygous carriers show two peaks with wild type probe as well as mutant probe. The homozygous normal shows a single peak in all cases. With the wild type probes (A, C), the mutant peak shifts to the left, whereas the mutant probes show a shift to the left of the wild type peak (B, D)

shows that a combination of wild-type and mutant probes may be more specific in determining the presence or absence of a known mutation. If the technique is used in preimplantation diagnosis to detect mutations in single cells from embryos prior to implantation, it is most important to assess cells that do not contain the mutation. Possible false positives due to a PCR mutation in the probe-binding area in the very first cycle of PCR are no problem, because only negative embryos are used for implantation.

The whole procedure, from selecting the cells until the final results takes less than 1 h. Consequently, the technique has great potential in fields where reliable and extremely fast mutation detection is important, such as preimplantation diagnosis.

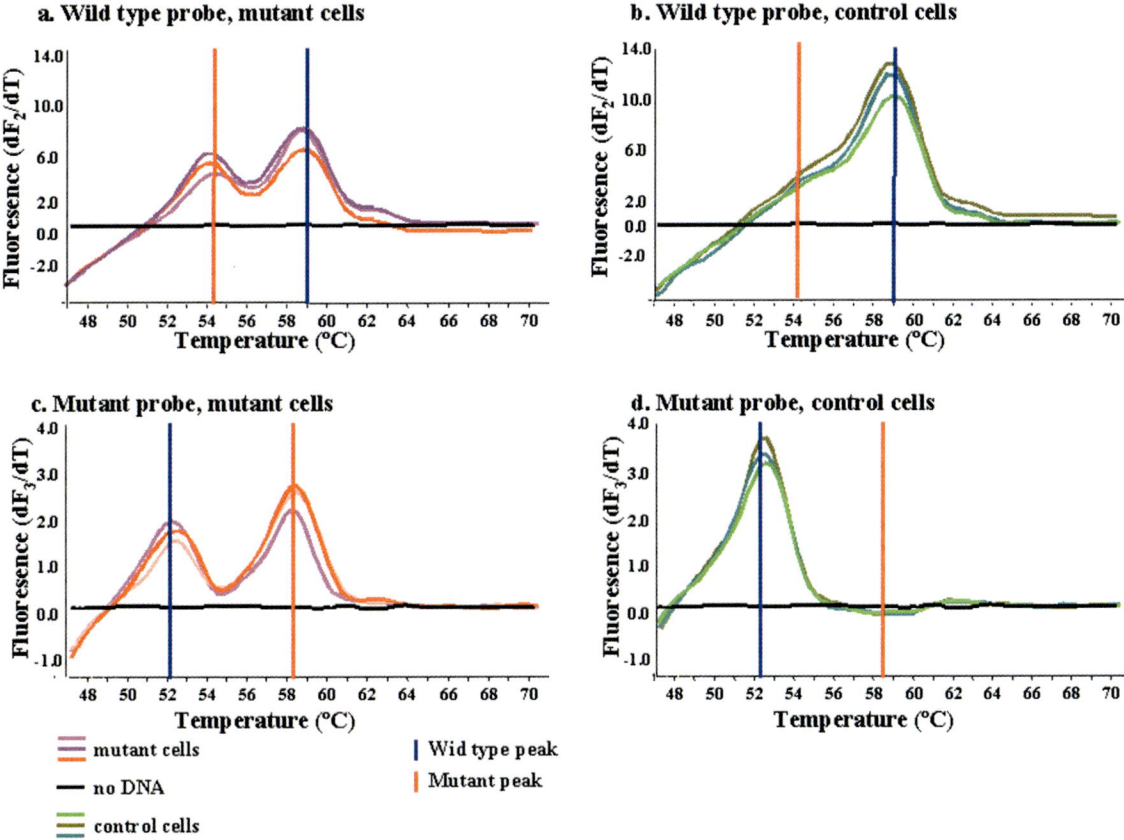

Fig. 2a–d. Detection of the *BRCA1 2841GT* mutation in single cells with wild-type probe (**a, b**) and mutant probe (**c, d**). The cells were from cell lines from a heterozygote carrier of the mutation (**a, c**) and a normal control (**b, d**). Each graph contains three lines from three separate cells. The *black line* at the *bottom* is the no DNA (no cell) control. Mutant and wild-type probe were used in a single tube. The heterozygous cells clearly show two distinct melting peaks with both probes (**a, b**). The wild-type cells show a single peak at 53°C with the mutant probe (**d**) and at 59°C with the wild-type probe (**b**)

Mutation Detection in Microdissected Tumor Samples

Presence of the *BRCA1 3519G→T* mutation was assessed in DNA from a small number (10–50) of normal and tumor cells from paraffin-embedded tissue from a carrier of this mutation. Two tumors were studied. The normal tissue showed a double peak in the melting curve, in keeping with heterozygosity for the mutation. The tumor DNA showed a single mutant peak (Fig. 3), indicating loss of the normal allele. Since *BRCA1* is considered a tumor suppressor gene, loss of the normal allele is expected in tumor tissue. With the present technique it is possible to detect loss of the normal allele in very small foci of aberrant tissue.

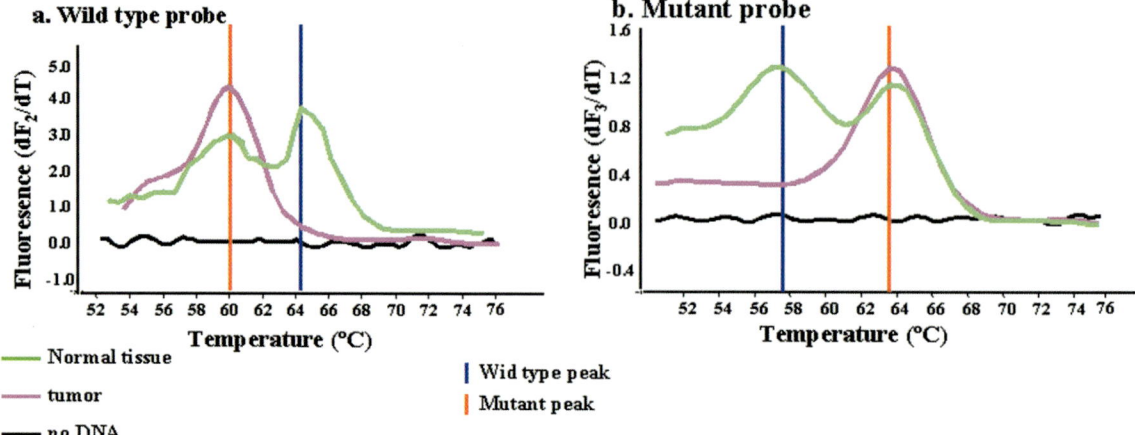

Fig. 3. Detection of the *BRCA1 3519G→T* mutation in microdissected tumor and normal tissue from a carrier of this mutation. As expected, normal tissue showed two melting peaks and is heterozygote for the mutation. The tumor tissue, however, shows only the mutant melting peak at 60.2 (wild type probe) or 63.7 °C (mutant probe).

References

1. Pals G, Young C, Mao HS, Worsham MJ (2001) Detection of a single base substitution in a single cell using the LightCycler. J Biochem Biophys Methods 47:121–129
2. Peelen T, van Vliet M, Petrij-Bosch A et al. (1997) A high proportion of novel mutations in BRCA1 with strong founder effects among Dutch and Belgian hereditary breast and ovarian cancer families. Am J Hum Genet 60:1041–1049
3. http://www.nhgri.nih.gov/Intramural_research/Lab_transfer/Bic/
4. Pals G, Pindolia K, Worsham MJ (1999) A rapid and sensitive approach to mutation detection using real-time polymerase chain reaction and melting curve analyses, using BRCA1 as an example. Mol Diagn 4:241–246

Rapid Screening for Five Major Cystic Fibrosis Mutations by Melting Peak Analysis Using Fluorogenic Hybridization Probes

SIEGFRIED BURGGRAF*, NAEEM MALIK,
EDITH SCHUHMACHER, BERNHARD OLGEMÖLLER

Introduction

Cystic fibrosis (CF), also known as mucoviscidosis, is a disorder of the exocrine glands [1]. Symptoms of CF are secretion of abnormally viscous mucus and elevated sweat electrolytes. The accumulation of thick mucus in the epithelium of the respiratory system and digestive tract causes progressive respiratory and gastrointestinal problems, including liver disease and diabetes mellitus, resulting in high morbidity and reduced life expectancy of CF patients. Variations in the cystic fibrosis transmembrane conductance regulator (CFTR), a cyclic adenosine monophosphate regulated chloride channel, are responsible for CF [2]. CF is one of the most common autosomal single gene disorders in Caucasian populations. The *CFTR* gene has a size of 250,000 bp separated into 27 exons, most of which are not larger than 200 bp [3]. So far, almost 1,000 CF-causing mutations, more or less scattered over the whole *CFTR* gene, have been registered by the Cystic Fibrosis Genetic Analysis Consortium (http://www.genet.sickkids.on.ca/cftr/). Only persons with homozygote or compound heterozygote *CFTR* gene mutations suffer from CF. The prevalence of CF in North America and Europe is between 1/2,400 and 1/1,600 and 4%–5% of the population are healthy carriers of one CF mutation.

The size of the gene and the large number of mutations make genetic CF screening complicated. However, the majority of the genetic variations are extremely rare. The most frequent genetic variation (e.g., in 77% of the CF cases in Bavaria, Germany [4, 5]) is a 3-bp deletion encoding phenylalanine at residue 508 (ΔF508). The frequency of specific mutations varies between different ethnic groups. Usually, however, only four to five further mutant alleles occur at a relative frequency of around 1%, whereas for all other mutations the frequency is significantly lower. A screening for the five or six most frequent mutations thus results in a sensitivity of about 95%. This means that in CF patients, at least one of the mutated alleles is detected by the test. Adding more mutations, e.g., the 25 next frequent mutations worldwide, to the screening program only marginally improves this sensitivity.

Since there is strong evidence that CF patients benefit from early detection and treatment [6], a neonatal screening for CF is already performed in several countries.

---

* Siegfried Burggraf (✉) (e-mail: burggraf@labor-bo.de)
Labor Becker, Olgemöller und Kollegen, Führichstrasse 70, 81671 Munich, Germany

In the first month of life, CF babies show increased concentrations of immunoreactive trypsinogen (IRT). IRT is measured from dried blood spot samples (Guthrie cards) collected from babies 3 days after birth. However, the specificity of this test in the first week of life is very low. Therefore, blood spots exhibiting elevated IRT are tested for common CF mutations. This so-called two-tiered IRT/DNA strategy has been shown to result in a good positive predictive value for CF [7, 8].

Here, we describe a procedure to rapidly detect five common CF mutations from dried blood spots using fluorogenic hybridization probes and dual color melting point analysis.

Materials

Equipment LightCycler instrument

Reagents QIAamp DNA Blood Mini Kit (Qiagen, Hilden, Germany)
Proteinase K (Qiagen)
Amplification Primer (TIB MOLBIOL, Berlin, Germany; GENSET, Paris, France)
Hybridization Probes (TIB MOLBIOL; GENSET)
LightCycler FastStart DNA Master Hybridization Probes (Roche Diagnostics, Mannheim, Germany)
DNAs with different CF mutations were kindly provided by T. Meitinger (Technische Universität München, Germany)

Procedure

Sample Preparation For DNA extraction, the QIAamp DNA Blood Mini Kit (Qiagen) was used. Two spots with a diameter of 3 mm were punched out from dried blood spots on Guthrie filter paper cards. In total, 180 µl of buffer ATL (Qiagen) was added to the spots. After incubation for 10 min at 85°C, 25 µl of Proteinase K (>600 mAU/ml) was added, followed by incubation for 60 min at 56°C. The remaining steps in DNA purification were performed according to the manufacturer's instructions. DNA was eluted in 50 µl 5 mmol/l Tris (pH 7.5).

Primer and Probe Design Since the test is planned to be used in a CF newborn screening program in Bavaria, Germany, the five most frequent Bavarian CF mutations, i.e., *ΔF508, R347P, G542X, G551D*, and *R553X* were chosen [5]. The oligonucleotides used for the detection of different CF alleles are shown in Table 1. Two primer pairs are necessary to amplify the *R347P* and the *ΔF508* loci in the *CFTR* gene. The amplification results in PCR products of 187 bp and 218 bp, respectively. The hybridization probe systems for melting point analysis of the two loci are labelled with different dyes, i.e., LCRed640 and LCRed705, to allow multiplex detection in a single reaction. The R347P detection probe contains a mismatch to avoid probe dimerization. A single primer pair is designed to amplify a 240-bp *CFTR* gene region encompassing the *G542X, G551D*, and *R553X* mutations. Two hybridization probe systems, again labelled with different dyes, are used for multiplex detection. All detection probes are homologous to the wild-type sequence. When hybridized to the template, the distance between the two fluorophores attached to the probes is 2 bp for all probe systems.

LightCycler PCR

The following master mix was used to detect CF mutations $\Delta F508$ and $R347P$ with hybridization probes:

	Volume [µl]	[Final]
LightCycler FastStart DNA Master Hybridization Probes	2	1×
MgCl$_2$ (25 mM)	2.4	4.0 mM
Primers (10 µM)	1+1+1+1	0.5 µM
Probes (10 µM)	0.4+0.4+0.4+0.4	0.2 µM
H$_2$O (PCR grade)	5	
Total volume	15	

The following master mix was used to detect CF mutations $G542X$, $G551D$, and $R553X$ with hybridization probes:

	Volume [µl]	[Final]
LightCycler FastStart DNA Master Hybridization Probes	2	1×
MgCl$_2$ (25 mM)	3.4	5.25 mM
Primers (10 µM)	2.5+2.5	1.25 µM
Probes (10 µM)	0.4+0.4+0.4+0.4	0.2 µM
H$_2$O (PCR grade)	3	
Total volume	15	

A total of 15 µl master mix and 5 µl of DNA (concentration not determined) were added to each glass capillary placed in precooled adaptors. Sealed capillaries were centrifuged with the adaptors (1000 g for 1 min) and placed into the Light-Cycler rotor.

The following PCR protocol was used for amplification:
- Denaturation of DNA and activation of FastStart polymerase for 10 min at 95°C
- Amplification

Parameter	Value		
Cycles	35		
Type	Quantification		
	Segment 1	Segment 2	Segment 3
Target temperature [°C]	95	55	72
Incubation time [s]	0	10	30
Temperature transition rate [°C/s]	20	20	20
Acquisition Mode	None	Single	None
Gains	F1=1; F2=10; F3=80		

Table 1. Primer sequences and hybridization probes for detection of five CF mutations

CFTR (GenBank Accession #HSAC000111)				
	Position	Length	GC (%)	T_m (°C)
R347P primers				
CAGCCTTCTTCTTCTCAGGGT	79860	21	52.4	66.96
TTGTTTATTGCTCCAAGA	80046R	18	33.3	55.23
Product	79860–80046	187		
R347P probes				
TGACCGCCATGCGT*AGAAC-F	79988R	19	57.9	64.54
LCRed640-TGCAGAATGAGATGGTGGTG AATATTTTCCGGA-P	79967R	33	42.4	70.88
ΔF508 primers				
GTGATTTGATAATGACCTAATAATGATGG	99129	29	31	63.25
GCTTTGATGACGCTTCTGTATCTATATTC	99346R	29	37.9	65.94
Product	99129–99346	218		
ΔF508 probes[a]				
CAGTTTTCCTGGATTATGCCTGGCACCATT-F	99284	30	46.7	73.06
LCRed705-AGAAAATATCATCTTTGGTGTTT CCTATGA-P	99316	30	30	65.61
G542X/G551D/R553X primers				
ATATGATTACATTAGAAGGAAGATGTGCC	127357	29	34.5	67.16
TTTACATGAATGACATTTACAGCAAATG	127596R	28	28.6	65.70
Product	127357–127596	240		
G542X probes[a]				
GATTCCACCTTCTCCAAGAACTAT-F	127499R	24	41.7	65.78
LCRed705-TGTCTTTCTCTGCAAACTTGG AGATGTCCTATTACCAA-P	127473R	38	39.5	74.53
G551D/R553X probes[a]				
GTGGAGGTCAACGAGCAAGAA-F	127507	21	52.4	68.59
LCRed640-TCTTTAGCAAGGTGAATAACT AATTATTGGTCTAGCAAG-P	127530	39	33.3	71.36

* A deliberately introduced mismatch in the *R347P* detection probe.
[a] Variable nucleotides in the detection probes are underlined.

- Melting Curve Analysis

Parameter	Value		
Cycles	1		
Type	Melting curve		
	Segment 1	Segment 2	Segment 3
Target temperature [°C]	95	40	75
Incubation time [s]	10	120	0
Temperature transition rate [°C/s]	20	20	0.1
Acquisition Mode	None	None	Cont.
Gains	F1=1; F2=10; F3=80		

Results

Figure 1 shows the results of a CF genotyping experiment with four different DNA samples.

Agarose gel electrophoresis of products of *ΔF508/R347P* multiplex PCR results in two bands of 187 bp and 218 bp (Fig. 1a). Fluorescence monitoring of the hybridization probes for *R347P* detection is shown in Fig. 1b. Samples 3–5 exhibit one wild-type peak at 62°C. Sample 2 is heterozygous for *R347P* with the wild-type peak at 62°C and a mutation peak at 51°C. Figure 1c shows the melting point analysis of the *ΔF508* deletion locus. Samples 2 and 3 are heterozygous for the deletion (peaks at 66°C and 57°C, respectively). The other samples exhibit only the wild-type peak at 66°C (Table 2).

Only one band (240 bp) is generated with the *G542X/G551D/R553X* PCR by agarose gel electrophoresis (Fig. 1d). The melting point analysis of the *G551D/R553X* locus is shown in Fig. 1e. Samples 2 and 3 exhibit one wild-type peak at 65°C. Sample 5, which is heterozygous for *R553X*, shows two peaks: the wild-type peak at 65°C and a mutation peak at 59°C. Sample 4 is heterozygous for G551D; however, due to the small T_m difference between wild type and mutation, it is difficult to obtain a separation of the two melting peaks. Only one broad peak is visible at 62°C. From this analysis it is not possible to conclude whether this sample is heterozygous or homozygous for the *G551D* mutation. Figure 1f shows the detection of the G542X mutation. Sample 3 is heterozygous for this mutation, exhibiting peaks at 63.5°C and 57°C for wild type and mutation, respectively. All other samples are wild type (one peak at 63.5°C). The genotypes of the samples used in this study are summarized in Table 3.

Comments

Neonatal CF screening from Guthrie card blood samples by measuring IRT shows only a very low predictive value. Many unaffected babies also have high IRT values. Another problem is the contamination of samples with feces, resulting in erroneously elevated IRT values. A second test is therefore necessary to confirm CF. Genetic screening appears to be the most specific procedure. The high number of tests in a screening program requires a fast, safe, and easy-to-interpret test. A test based on melting point analysis of fluorescently labelled probes after high speed PCR amplification on a LightCycler instrument definitely meets these requirements. After nucleic acid extraction from the blood spots, five major cystic fibrosis mutations can be detected in less than 1 h.

The detection of a specific mutation should be verified by a separate test using a mutation-specific probe since there are several other, though extremely rare, CF-causing mutations, which also can cause a T_m shift in the melting analysis using the described probes. For example, *ΔI507* is detected with the *ΔF508* hybridization probes and shows the same T_m shift as a *ΔF508* deletion.

Running the entire test in a single capillary could be a goal for future progress in the technique. With the present probe design the probes labelled with the same

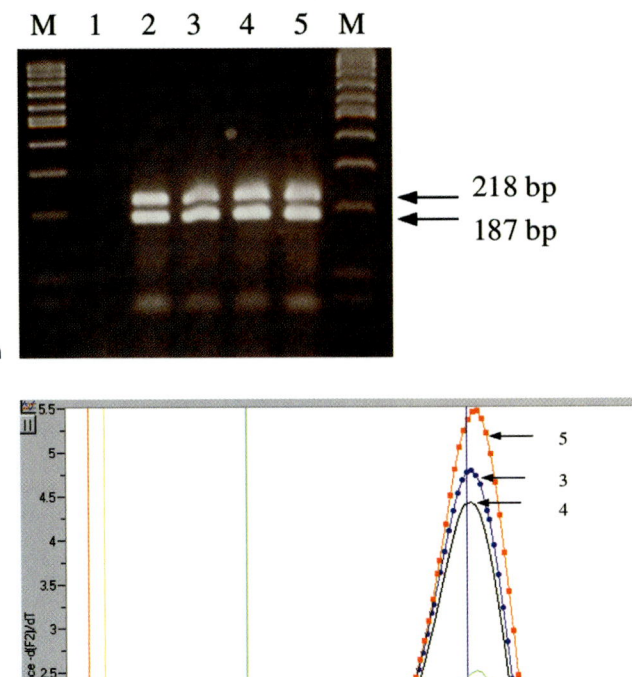

Fig. 1a–f. Agarose gel electrophoresis and melting peak analysis of PCR products from four different DNA samples and a negative control. The assignment of the samples to the melting peaks and the lanes in the agarose gel are shown by *numbers*. The result of the multiplex PCR for CF mutation loci *R347P* and *ΔF508* is shown in **a–c**. PCR of a region of the *CFTR* gene encompassing CF mutations *G542X*, *G551D*, and *R553X* is shown in **d–f**. Melting curves were converted to melting peaks by plotting the negative derivative of the fluorescence with respect to temperature (–dF/dT) against temperature. The melting point analyses of probes labelled with LCRed640 (probes for detection of *R347P* and *G551D/R553X*, respectively) are shown in **b** and **e**. The melting point analyses of probes labelled with LCRed705 (probes for detection of *ΔF508* and *G542X*, respectively) are shown in **e** and **f**. PCR products removed from glass capillaries by centrifugation were separated on 3% agarose gels. *M*, 100 bp molecular weight marker. *1*, negative control (H$_2$O); *2*, *R347P* and *ΔF508* heterozygous; *G542X*, *G551D*, and *R553X* wild type; *3*, *G542X* and *ΔF508* heterozygous; *R347P*, *G551D*, and *R553X* wild type; *4*, *G551D* heterozygous; *R347P*, *ΔF508*, *G542X*, and *R553X* wild type; *5*, *R553X* heterozygous; *R347P*, *ΔF508*, *G542X*, and *G551D* wild type

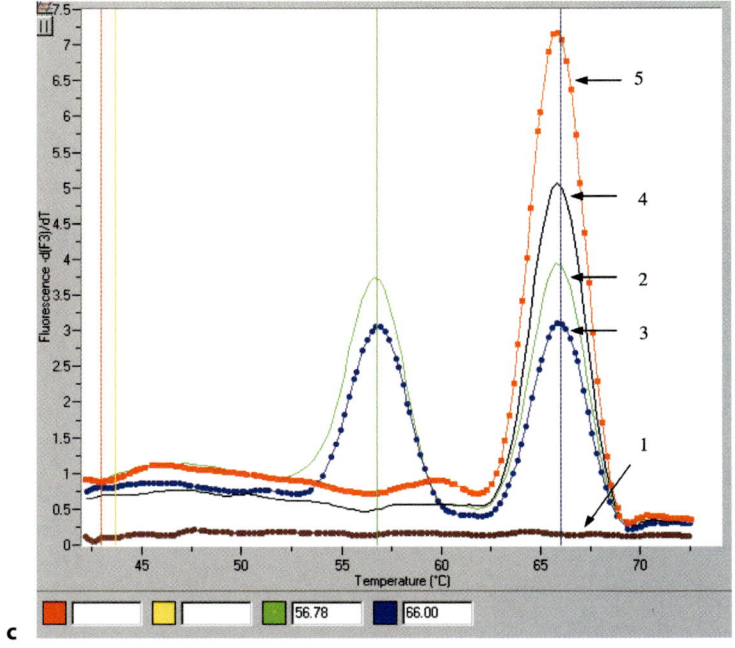

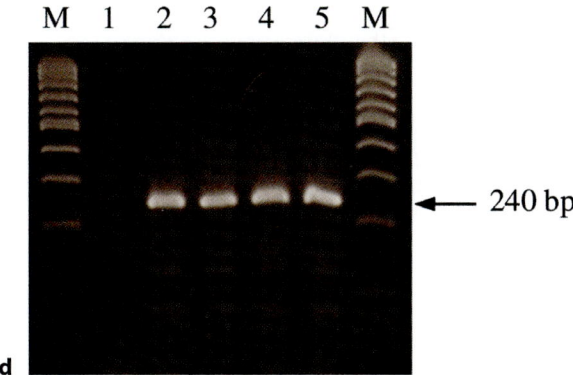

Fig. 1c, d.

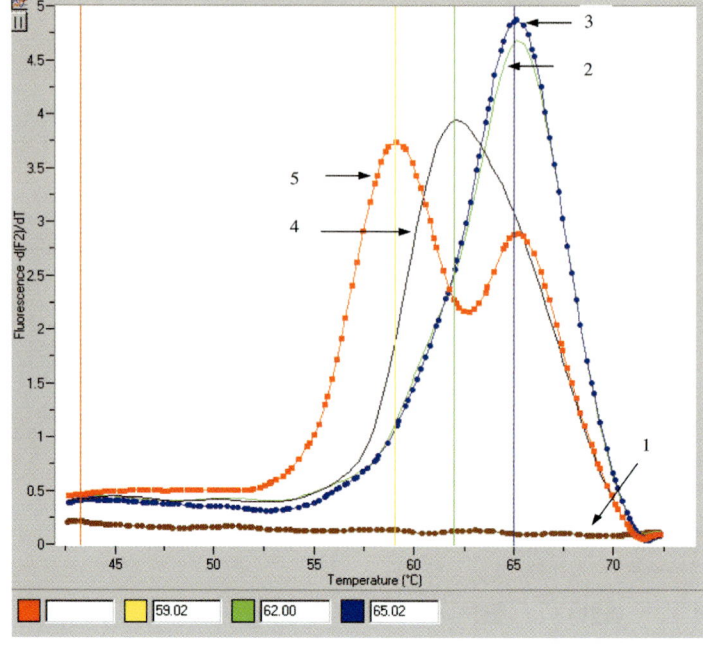

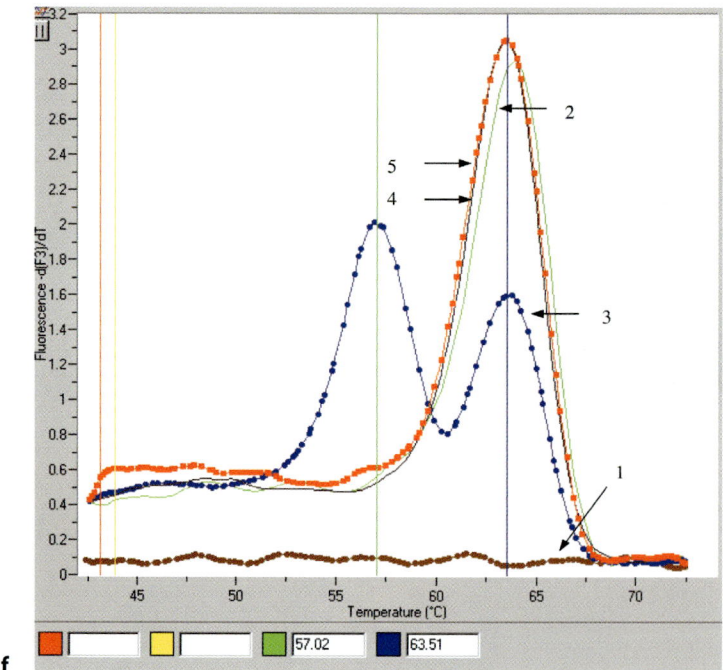

Fig. 1e, f.

Table 2. Melting temperatures of different alleles of the *CFTR* gene

Locus	Allele	Pairing	T_m (observed)
R347P	WT	C–G match	62°C
	Mutation	C–C mismatch	51°C
ΔF508	WT	TTT match	66°C
	Mutation	- - - Deletion	57°C
G542X	WT	C–G match	63.5°C
	Mutation	C–T mismatch	57°C
G551D	WT	C–G match	65°C
	Mutation	C–A mismatch	62°C
R553X	WT	G–C match	65°C
	Mutation	G–T mismatch	59°C

WT, wild-type sequence.

Table 3. Genotypes of the DNA samples used for CF mutation analysis

Sample	Mutation				
	R347P	*ΔF508*	*G542X*	*G551D*	*R553X*
1	–	–	–	–	–
2	HE	HE	WT	WT	WT
3	WT	HE	HE	WT	WT
4	WT	WT	WT	HE	WT
5	WT	WT	WT	WT	HE

WT, wild-type sequence; HE, heterozygous mutation.

dye, e.g., for *ΔF508* and *G542X* detection, show very similar T_ms. Therefore it is necessary to run the test in two separate capillaries. It may be possible to redesign the probes in order to obtain distinguishable T_ms for the probes labelled with the same dye.

References

1. Farber S (1945) Some organic digestive disturbances in early life. Pathological changes associated with pancreatic insufficiency. Mich Med Soc 44:587–594
2. Widdicombe JH, Welsh MJ, Finkbeiner WE (1985) Cystic fibrosis decreases the apical membrane chloride permeability of monolayers cultured from cells of tracheal epithelium. Proc Natl Acad Sci USA 82:6167–6171
3. Kerem B, Rommens JM, Buchanan JA, Markiewicz D, Cox TK, Chakravarti A, Buchwald M, Tsui LC (1989) Identification of the cystic fibrosis gene: genetic analysis. Science 245:1073–1080

4. Deufel T, Rabe H, Wieser T, Meitinger T, Rosenecker J, Bertele-Harms R, Harms K, Hadorn HB, Roscher AA (1993) Mutation analysis in the diagnosis of cystic fibrosis. Eur J Pediatr 152:909–911
5. Tümmler B, Storrs T, Dziadek V. et al. (1996) Geographic distribution and origin of CFTR mutations in Germany. Hum Genet 97:727–731
6. Dankert-Roelse JE, te Meerman GJ (1995) Long term prognosis of patients with cystic fibrosis in relation to early detection by neonatal screening and treatment in a cystic fibrosis centre. Thorax 50:712–718
7. Ranieri E, Lewis BD, Gerace RL, Ryall RG, Morris CP, Nelson PV, Carey WF, Robertson EF (1994) Neonatal screening for cystic fibrosis using immunoreactive trypsinogen and direct gene analysis: four years' experience. BMJ 308:1469–1472
8. Murray J, Cuckle H, Taylor G, Littlewood J, Hewison J (1999) Screening for cystic fibrosis. Health Technol Assess 3:1–104

LightCycler PCR for the Polymorphisms −308 and −238 in the *TNF Alpha* Gene and for the TNFB1/B2 Polymorphism in the *LT Alpha* Gene

Lukas Bestmann, Näder Helmy, Felicia Garofalo, Aynur Demirtas, Dieter Vonderschmitt, Friedrich E. Maly*

Introduction

Tumor necrosis factor alpha (TNF-α), a multifunctional cytokine of ~17 kDa composed of 157 amino acids, was initially identified as a macrophage-derived serum protein mediating necrosis of solid tumors in mice and lysis of several types of transformed cells in vitro. Indeed, recombinant tumor necrosis factor alpha (rTNF-α) has potent antitumor activity in experimental studies on human tumor xenografts. However, in humans, the administration of rTNF-α is hampered by severe systemic side effects, with the maximum tolerated dose being at least tenfold less than the effective dose in animals [14]. Thus, systemic treatment with TNF has not become a viable treatment modality for common human cancers. Isolated perfusion of the limbs allows the delivery of high-dose rTNF-α in a closed system with acceptable side effects, and several studies have shown that TNF-α, in combination with other agents such as melphalan, does possess activity against certain human tumors in vivo [12, 20].

Apart from tumor biology, TNF-α is a major mediator of inflammation and infection. Although also required for normal immune responses, the overexpression of TNF-α has severe pathological consequences. For example, it is responsible for some of the severe effects during Gram-negative sepsis and plasma levels of TNF-α are positively correlated with severity and mortality in various illnesses such as malaria and leishmaniasis [4, 18]. Furthermore, it is the major mediator of cachexia observed in tumor patients and congestive heart failure, hence its earlier name Cachectin.

Tumor necrosis factor beta (TNF-β) or Lymphotoxin-α, a glycoprotein of 20–25 kDa with antiviral and antitumor activities, is produced by activated T and B lymphocytes and shares sequence homologies and some biologic actions (such as in vitro cytotoxicity against certain tumor cells) with TNF-α. Contrary to TNF-α, TNF-β is barely detectable in acute inflammation and sepsis [25], but forms membrane-associated heteromeric complexes with Lymphotoxin-β and has a central role in the development of lymphoid organs and in chronic inflammation. Lymphoid organ development and inflammation have previously been consid-

---

* Friedrich E. Maly (✉) (e-mail: fma@ikc.unizh.ch)
  Institute of Clinical Chemistry, University Hospital Zurich, Rämistrasse 100, 8091 Zürich, Switzerland

ered as mechanically and functionally distinct. In recent years, it has been realized that these phenomena have much in common and that Lymphotoxin-α, Lymphotoxin-β, TNF-α, and their multiple receptors participate in lymphoid organ development and chronic inflammation [23].

TNF-α and TNF-β bind to members of the same TNF receptor family, however, not on the same binding domains and thus cause partially comparable but also partially different biological effects [27].

The *TNF-α* and *TNF-β* genes are located in tandem on human chromosome *6p21.3* within the class III region gene cluster of the major histocompatibility complex (*MHC*), ≈250 kb centromeric of the *HLA-B* locus and 850 kb telomeric of *HLA-DR*. Their high homology and close location in the genome, as well as evolutionary studies, suggest a common ancestor for both genes that duplicated during evolution [19].

Several polymorphisms have so far been noted in TNF-α and TNF-β genes (LT-α gene). Two promoter polymorphisms in the *TNF-α* gene are relatively frequent: G-238A at –238 relative to the transcriptional start site is found in about 5%, G-308A at –308 relative to the transcriptional start site in about 30% of Caucasians, and both modulate in vitro production of TNF-α by mononuclear cells (NCBI GenBank Accession #L11698). The *–308A* allele of the promoter polymorphism is part of the extended haplotype *HLA-A1-B8-DR3-DQ2*, which is associated with autoimmunity and high TNF production. The TNF-β A329G polymorphism, identical to the NcoI RFLP in intron 1 (NCBI GenBank Accession #M55913; NcoI will cut at position 324), is even more common (40%–45%) [1]. Clinically, these polymorphisms have so far been associated with septic shock and survival of sepsis, various autoimmune diseases (insulin-dependent diabetes mellitus type I [3] and lupus erythematosus [21], Morbus Crohn, rheumatoid arthritis [2]), atopy [5] and asthma [13], an increased risk of rejection after kidney transplantation and the response to pathogens such as hepatitis B and C viruses [9, 10], and chlamydia trachomatis.

In our study we examined correlations of these TNF polymorphisms with the risk of developing sepsis after severe polytrauma. We found the –238A variant of the *TNF-α* gene associated with sepsis development, but no correlation with the G-308A polymorphism of the same gene and with A329G in the *TNF-β* gene [8], using the LightCycler to detect these polymorphisms.

The methods used are described in detail in the following section.

Materials

Equipment LightCycler instrument (Roche Diagnostics, Mannheim, Germany)
Oligo Primer Analysis Software (MedProbe, Oslo, Norway)

Reagents Amplification Primers and Hybridization Probes combined in ToolSets (Genes-4U, Neftenbach, Switzerland, www.Genes-4U.com)
Qiagen QIAmp Blood kit (Hilden, Germany)
LightCycler-DNA Master Hybridization Probes (Roche Diagnostics, Mannheim, Germany)

Procedure

DNA was isolated by a standard rapid lysis technique using the Qiagen QIAmp Blood kit.

Preparation of Template DNA

For the *TNF-α* mutation analysis, we used complementary designed sensor probes, fitting to the wild type, covering the point mutations at position –308 and –238 (GenBank Accession #L11698). The anchor probe was designed complementary to an adjacent invariant gene segment. Thus, the wild-type allele (GG) is expected to yield a higher melting temperature than the *–308* mutation (AA), measured on channel two. The *–238* mutant shows the same behaviour, except the measurement is on channel three. The anchor probe was modelled between the two reporter probes, capable of detecting both mutations in the same run. PCR was performed with the nucleotides mentioned in Table 1.

For the TNF-β assay (GenBank Accession #M55913), the probes were designed in a similar manner. Because of contradictory statements in literature concerning the designation of the wild-type and the mutated allele, the neutral nucleotide

Primer and Probe Design

Table 1. Oligonucleotides

TNF-α –308 and –238 (GenBank Accession #L11698)				
	Position	Length	GC (%)	T_m (°C)
Primers				
5'-TTC CTG CAT CCT GTC TGG AA	658	20	50.0	56.8
5'-CAG CGG AAA ACT TCC TTG GT	966–985	20	50.0	58.7
Product	658–985	328 bp		
Probes				
5'-AAT AGG TTT TGA GGG GCA TGG GGAC T-LCRed640-P	746	25	48.0	65.9
5'-LCRed705-CCT CGG AAT CGG AGC AGG GAG GA-P	816	23	65.2	68.6
5'-F-TTC AGC CTC CAG GGT CCT ACA CAC AAA TCA GTC AGT GGC CCA GAA GA-F	755	47	53.2	81.0
TNF-β A329G (GenBank Accession #M55913)				
Primers				
5'-CCT GCA CCT GCT GCC TGG AT	193	20	65.0	64.8
5'-CAG TCA GAG AAA CCC CAA GGT GA	369	23	52.2	60.9
Product	193–392	200 bp		
Probes				
5'-LCRed640-TCT CTG TTT CTG CCA TGG TTC CTC TCT G P	312	28	50.0	66.0
5'-CTC CAT CTG TCA GTC TCA TTG TCT CTG TCA CAC A-F	277	34	47.1	66.6

nomenclature is used. The sensor probe is designed on the GG genotype. The anchor probe contains no mismatches.

LightCycler PCR

The following master mixes were used for amplification and hybridization based detection of the TNF gene polymorphisms (for clarity, individual ToolSet components are given):

TNF-α G-308A and G-238A	Volume [µl]	[Final]
Primers (10 µM each)	1.0 + 1.0	0.5 µM each
Hybridization probes (10 µM each)	0.4 + 0.4	0.2 µM each
LightCycler-DNA Master Hybridization Probes	2.0	1×
$MgCl_2$ stock solution (25 mM)	0.8	2.0 mM
H_2O (PCR grade)	12.0	–
Total volume	18.0	–

To complete the TNF-α G-308A and G-238A amplification mixture, 18 µl of master mix and 2 µl of isolated DNA solution were added to each capillary. After a short centrifugation, the sealed capillaries were placed into the LightCycler rotor. The inclusion of negative as well as positive controls in each set of experiments is considered to be obligatory in the field of diagnostic PCR. The negative control sample was prepared by replacing the DNA template with PCR-grade water or elution buffer, respectively.

Preparation of TNF-β master mix (for clarity, individual Tool Set components are given):

TNF-β A329G	Volume [µl]	[Final]
Primers (10 µM each)	1.0 + 1.0	0.5 µM each
Hybridization probes (10 µM each)	0.4 + 0.4	0.2 µM each
LightCycler-DNA Master Hybridization Probes	2.0	1×
$MgCl_2$ stock solution (25 mM)	0.8	2.0 mM
H_2O (PCR grade)	10.4	–
Total volume	16.0	–

To complete the TNF-β A329G amplification mixture, 16 µl of master mix and 4 µl of isolated DNA solution were added to each capillary. After a short centrifugation, the sealed capillaries were placed into the LightCycler rotor. Again, negative as well as positive controls were included in each set of experiments. The negative control sample was prepared by replacing the DNA template with PCR-grade water or elution buffer, respectively.

LightCycler settings for TNF-α G-308A and G-238A:

Denaturation

Cycles	1				
Type	None				
Segment Number	Temperature Target [°C]	Hold Time [s]	Slope [°C/s]	Acquisition Mode	Fluorescence Display Mode
1	95	120	20	None	F2/1 (G-308A)
					F3/1 (G-238A)

Amplification

Cycles	70				
Type	None				
Segment Number	Temperature Target [°C]	Hold Time [s]	Slope [°C/s]	Acquisition Mode	Fluorescence Display Mode
1	95	0	20	None	F2/1 (G-308A)
2	50	10	20	Single	F3/1 (G-238A)
3	72	10	5	None	

Melting

Cycles	1				
Type	Melting Curves				
Segment Number	Temperature Target [°C]	Hold Time [s]	Slope [°C/s]	Acquisition Mode	Fluorescence Display Mode
1	95	0	20	None	F2/1 (G-308A)
2	45	90	20	None	F3/1 (G-238A)
3	85	0	0.1	Continuous	

Cooling

Cycles	1				
Type	None				
Segment Number	Temperature Target [°C]	Hold Time [s]	Slope [°C/s]	Acquisition Mode	Fluorescence Display Mode
1	40	30	20	None	F2/1 (G-308A)
					F3/1 (G-238A)

Fluorescence settings

LED power	CALIB
F1 Gain	1
F2 Gain	15
F3 Gain	30

Experimental Protocols

LightCycler settings for TNF-β A329G:

Denaturation

Cycles	1				
Type	None				
Segment Number	Temperature Target [°C]	Hold Time [s]	Slope [°C/s]	Acquisition Mode	Fluorescence Display Mode
1	95	0	0	None	F2/1 (A329G)

Amplification

Cycles	45				
Type	None				
Segment Number	Temperature Target [°C]	Hold Time [s]	Slope [°C/s]	Acquisition Mode	Fluorescence Display Mode
1	95	0	20	None	F2/1 (A329G)
2	60	10	20	Single	
3	72	13	5	None	

Melting

Cycles	1				
Type	Melting Curves				
Segment Number	Temperature Target [°C]	Hold Time [s]	Slope [°C/s]	Acquisition Mode	Fluorescence Display Mode
1	95	0	20	None	F2/1 (A329G)
2	45	30	20	None	
3	80	0	0.1	Continuous	

Cooling

Cycles	1				
Type	None				
Segment Number	Temperature Target [°C]	Hold Time [s]	Slope [°C/s]	Acquisition Mode	Fluorescence Display Mode
1	40	30	20	None	F2/1 (A329G)

Fluorescence settings

LED power	CALIB
F1 Gain	1
F2 Gain	7
F3 Gain	10

Results

Based on the GenBank sequences of the TNF-α and TNF-β genes, we established PCR protocols for genotyping analysis for three common single nucleotide polymorphisms (SNPs) in the *TNF* gene cluster (Figs. 1–3). Genotype assignments were verified by sequencing on an ABI Prism 310 Sequencer (Applied Biosystems, Rotkreuz, Switzerland).

Comments

The TNF-α G-238A, TNF-α G-308A and LT-α/TNF-β NcoI polymorphisms (A329G) are frequent genetic alterations that have been linked to a great variety of disease states: various autoimmune diseases, atopy [28] and asthma [13], an increased risk of rejection after transplantation and the response to hepatitis B and C viruses [9, 10, 22] and chlamydia trachomatis [6]. Particularly, TNF-α G-238A appears to identify a small subgroup of severely polytraumatized patients likely to develop sepsis and may be used as a prognostic tool [26, 27].

As TNF is a major mediator in the pathogenesis of sepsis and its sequelae, individuals with a genetic determination for high TNF responses may be at high risk for development of organ failure and death when challenged with severe infec-

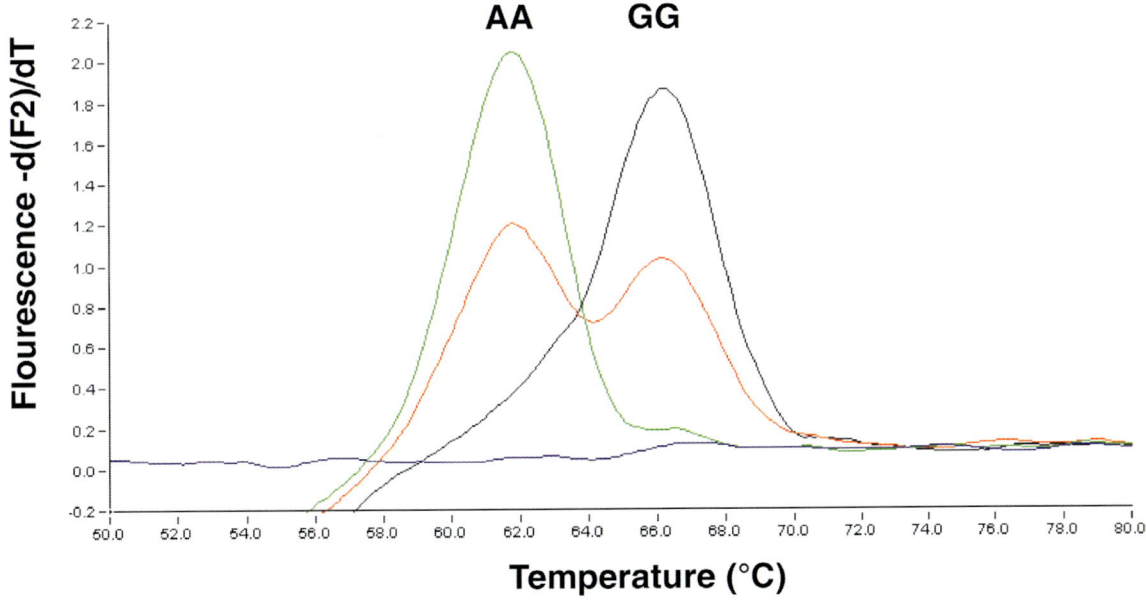

Fig. 1. Melting curve analysis of TNF-α –308 polymorphism. The *black (GG)* and the *green (AA) curves* show the homozygous variants of wild type and mutation, respectively. The *red curve* shows the heterozygous carrier of the mutation. The *blue base line* represents the buffer

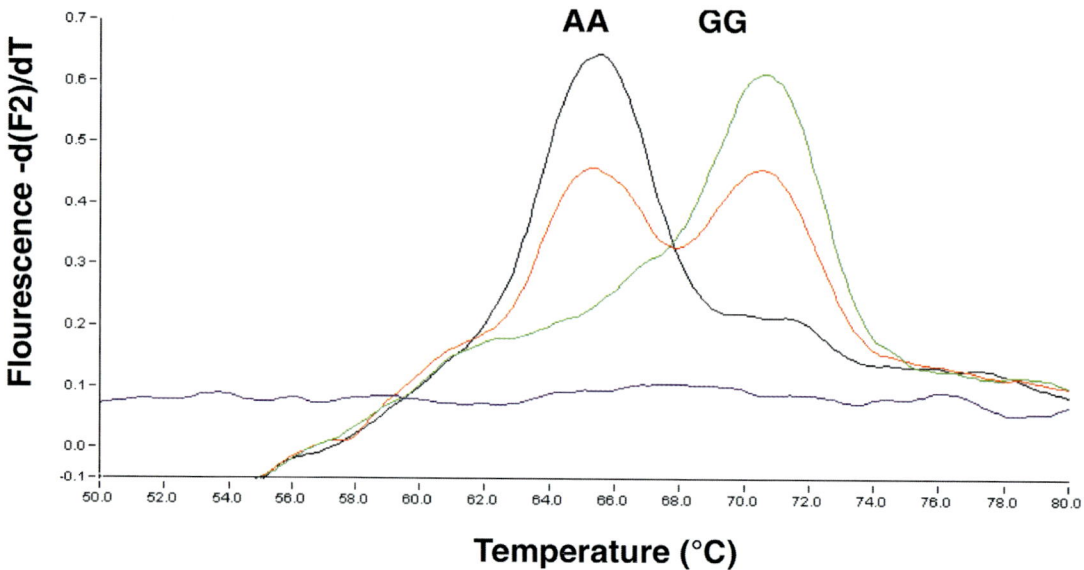

Fig. 2. Melting curve analysis of TNF-α –238 polymorphism. The *green (GG)* and the *black (AA) curves* show the homozygous variants of wild type and mutation, respectively. The *red curve* shows the heterozygous carrier of the mutation. The *blue base line* represents the buffer

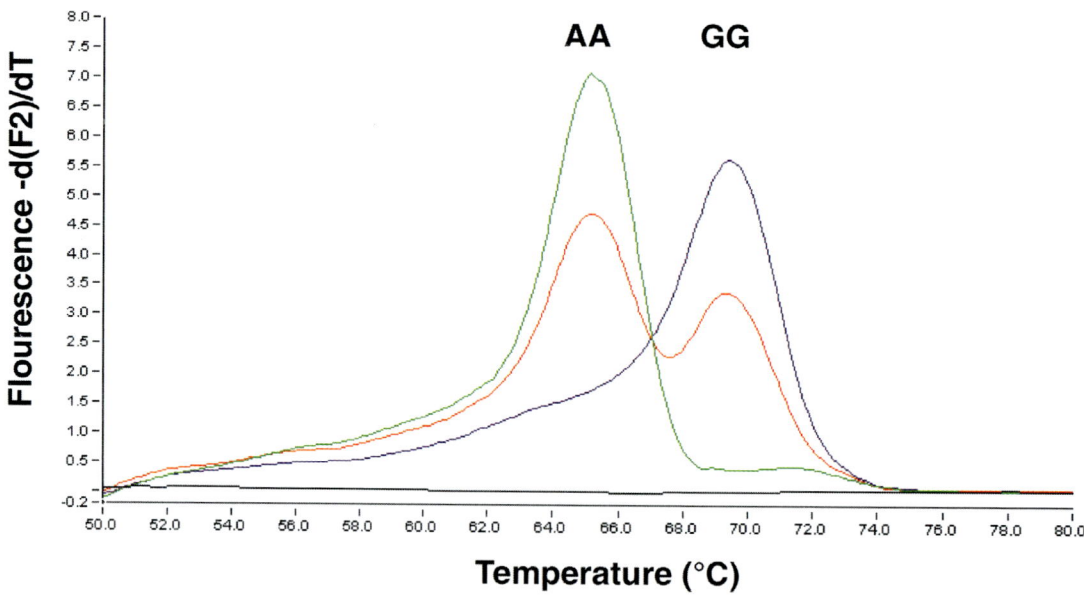

Fig. 3. Melting curve analysis of the TNF-β 329 polymorphism. The *blue (GG)* and the *green (AA) curves* show the homozygous variants of wild type and mutation, respectively. The *red curve* shows the heterozygous carrier of the mutation. The *black base line* represents the buffer

tion, trauma and other noxious stimuli that can provoke a generalized systemic inflammatory response [7]. It may thus become possible to target anti-TNF treatment specifically at those individuals genetically predisposed to produce high amounts of this mediator after polytrauma. TNF is also a critical mediator in rheumatoid arthritis and may therefore be a useful target for specific immunotherapy. Indeed, beneficial responses in patients have been observed after treatment with a chimeric monoclonal antibody to TNF (cA2) aiming to reduce the amount of bioactive TNF [2]. Again here, analysis of TNF polymorphisms may become useful in selecting those individuals most likely to benefit from this treatment. In recent developments, a critical role of the TNF system has also been highlighted for diseases caused by prions. For example, TNF-α-deficient mice have been shown resistant to peripheral infections with scrapie [16]. It is an exciting possibility that genetic variations at the TNF loci may modulate the natural history of prion infections such as bovine spongiform encephalopathy (BSE or mad cow disease) and the human new variant Creutzfeld-Jacob disease likely caused by the same agent. In this and other fields of clinical investigation, methods for determining TNF locus variants such as those described here may find widespread application in the near future.

References

1. Agarwal P, Oldenburg MC, Czarneski JE, Morse RM, Hameed MR, Cohen S, Fernandes H (2000) Comparison study for identifying promoter allelic polymorphism in interleukin 10 and tumor necrosis factor alpha genes. Diagn Mol Pathol 9:158–164
2. Baumgartner SW (2000) Tumor necrosis factor inactivation in the management of rheumatoid arthritis. South Med J 93:753–759
3. Braun J, Marz W, Winkelmann BR, Donner H, Henning Usadel K, Badenhoop K (1998) Tumour necrosis factor β alleles an hyperinsulinaemia in coronary artery disease. Eur J Clin Invest 28:538–542
4. Cabrera M, Shaw MA, Sharples C, Williams H, Castes M, Convit J, Blackwell JM (1995) Polymorphism in tumor necrosis factor genes associated with mucocutaneous leishmaniasis. J Exp Med 182:1259–1264
5. Castro J, Telleria JJ, Linares P, Blanco-Quiros A (2000) Increased TNFA*2, but not TNFB*1, allele frequency in Spanish atopic patients. J Investig Allergol Clin Immunol 10:149–154
6. Conway DJ, Holland MJ, Bailey RL, Campbell AE, Mahdi OS, Jennings R, Mbena E, Mabey DC (1997) Scarring trachoma is associated with polymorphism in the tumor necrosis factor alpha (TNF-alpha) gene promoter and with elevated TNF-alpha levels in tear fluid. Infect Immun 65:1003–1006
7. Flach R, Majetschak M, Heukamp T, Jennissen V, Flohe S, Borgermann J, Obertacke U, Schade FU (1999) Relation of ex vivo stimulated blood cytokine synthesis to post-traumatic sepsis. Cytokine 11:173–178
8. Helmy N, Bestmann L, Garofalo F, Maly FE (2000) Tumor necrosis factor alpha polymorphism G-238A is associated with sepsis after trauma (abstract). Shock 13 [Suppl]:486
9. Hohler T, Kruger A, Gerken G, Schneider PM, Meyer zum Buschenefelde KH, Rittner C (1998) A tumor necrosis factor-alpha (TNF-alpha) promoter polymorphism is associated with chronic hepatitis B infection. Clin Exp Immunol 111:579–582
10. Hohler T, Kruger A, Gerken G, Schneider PM, Meyer zum Buschenfelde KH, Rittner C (1998) Tumor necrosis factor alpha promoter polymorphism at position −238 is associated with chronic active hepatitis C infection. J Med Virol 54:173–177

11. Kaluza W, Reuss E, Grossmann S, Hug R, Schopf RE, Galle PR, Maerker-Hermann E, Hoehler T (2000) Different transcriptional activity and in vitro TNF-alpha production in psoriasis patients carrying the TNF-alpha 238A promoter polymorphism. J Invest Dermatol 114: 1180–1183
12. Lejeune F, Lienard D, Eggermont A, Schraffordt Koops H, Rosenkaimer F, Gerain J, Klaase J, Kroon B, Vanderveken J, Schmitz P (1995) Administration of high-dose tumor necrosis factor alpha by isolation perfusion of the limbs. Rationale and results. J Infus Chemother 5:73–81
13. Li Kam Wa TC, Mansur AH, Britton J, Williams G, Pavord I, Richards K, Campbell DA, Morton N, Holgate ST, Morrison JF (1999) Association between –308 tumour necrosis factor promoter polymorphism and bronchial hyperreactivity in asthma. Clin Exp Allergy 29: 1204–1208
14. Lienard D, Eggermont AM, Koops HS, Kroon B, Towse G, Hiemstra S, Schmitz P, Clarke J, Steinmann G, Rosenkaimer F, Lejeune FJ (1999) Isolated limb perfusion with tumour necrosis factor-alpha and melphalan with or without interferon-gamma for the treatment of in-transit melanoma metastases: a multicentre randomized phase II study. Melanoma Res 9:491–502
15. Louis E, Franchimont D, Piron A, Gevaert Y, Schaaf-Lafontaine N, Roland S, Mahieu P, Malaise M, De Groote D, Louis R, Belaiche J (1998) Tumour necrosis factor (TNF) gene polymorphism influences TNF-α production in lipopolysaccharide (LPS)-stimulated whole blood cell culture in healthy humans. Clin Exp Immunol 113:401–406
16. Mabbott NA, Williams A, Farquhar CF, Pasparakis M, Kollias G, Bruce ME (2000) Tumor necrosis factor alpha-deficient, but not interleukin-6-deficient, mice resist peripheral infection with scrapie. J Virol 74:3338–3344
17. McGarry F, Walker R, Sturrock R, Field M (1999) The –308.1 polymorphism in the promoter region of the tumor necrosis factor gene is associated with ankylosing spondylitis independent of HLA-B27. J Rheumatol 26:1110–1116
18. Mira JP, Cariou A, Grall F, Delclaux C, Losser MR, Heshmati F, Cheval C, Monchi M, Teboul JL, Riche F, Leleu G, Arbibe L, Mignon A, Delpech M, Dhainaut JF (1999) Association of TNF2, a TNF-alpha promoter polymorphism, with septic shock susceptibility and mortality: a multicenter study. JAMA 282:561–568
19. Nedwin GE, Naylor SL, Sakaguchi AY, Smith D, Jarrett-Nedwin J, Pennica D, Goeddel DV, Gray PW (1985) Human lymphotoxin and tumor necrosis factor genes: structure, homology and chromosomal localization. Nucleic Acids Res 13:6361–6373
20. Olieman AF, Lienard D, Eggermont AM, Kroon BB, Lejeune FJ, Hoekstra HJ, Koops HS (1999) Hyperthermic isolated limb perfusion with tumor necrosis factor alpha, interferon gamma, and melphalan for locally advanced nonmelanoma skin tumors of the extremities: a multicenter study. Arch Surg 134:303–307
21. Rood MJ, van Krugten MV, Zanelli E, van der Linden MW, Keijsers V, Schreuder GM, Verduyn W, Westendorp RG, de Vries RR, Breedveld FC, Verweij CL, Huizinga TW (2000) TNF-308A and HLA-DR3 alleles contribute independently to susceptibility to systemic lupus erythematosus. Arthritis Rheum 43:129–134
22. Rosen HR, Lentz JJ, Rose SL, Rabkin J, Corless CL, Taylor K, Chou S (1999) Donor polymorphism of tumor necrosis factor gene: relationship with variable severity of hepatitis C recurrence after liver transplantation. Transplantation 68:1898–1902
23. Ruddle NH (1999) Lymphoid neo-organogenesis: lymphotoxin's role in inflammation and development. Immunol Res 19:119–125
24. Skeie GO, Pandey JP, Aarli JA, Gilhus NE (1999) TNFA and TNFB polymorphisms in myasthenia gravis. Arch Neurol 56:457–461
25. Sriskandan S, Moyes D, Lemm G, Cohen J (1996) Lymphotoxin-alpha (TNF-beta) during sepsis. Cytokine. 8:933–937
26. Stuber F, Udalova IA, Book M, Drutskaya LN, Kuprash DV, Turetskaya RL, Schade FU, Nedospasov SA (1996) -308 tumor necrosis factor (TNF) polymorphism is not associated

with survival in severe sepsis and is unrelated to lipopolysaccharide inducibility of the human TNF promoter. J Inflamm 46:42–50
27. Stuebner F, Petersen M, Bokelmann F, Schade U (1996) A genomic polymorphism within the tumor necrosis factor locus influences plasma tumor necrosis factor-α concentrations and outcome of patients with severe sepsis. Crit Care Med 24:381–384
28. Trabetti E, Patuzzo C, Malerba G, Galavotti R, Martinati LC, Boner AL, Pignatti PF (1999) Association of a lymphotoxin alpha gene polymorphism and atopy in Italian families. J Med Genet 36:323–325

Rapid Genotyping of 2-bp and 9-bp Deletion Mutations Using the LightCycler Instrument

Tsutomu Aoshima, Mitsuharu Kajita, Yoshitaka Sekido, Shunji Mimura, Kazuyoshi Watanabe, Kaoru Shimokata, Toshimitsu Niwa*

Introduction

Instead of sequencing the candidate gene PCR fragment, simple methods such as restriction enzyme digestion after PCR are used to detect known mutation. In some cases, however, such methods cannot be applied and sometimes lead to an ambiguous result. A faster method would be required in case of mass screening or emergency. Recently, an elegant method to detect a known mutation with the LightCycler has been introduced. Rapid real-time PCR is monitored by fluorescent oligonucleotide probes that hybridize the target region of the PCR product. Only when the probes hybridize the template fluorescence resonance energy transfer occurs thereby producing a specific fluorescence emission. Then, by slowly heating the PCR product in the probe-hybridized state, the diminishing fluorescence value gives a melting curve. In the presence of mismatch sequences between the template and the probe, the probe melts off at a lower temperature. Therefore, the melting curve clearly demonstrates its genotype as the differences in T_m. This fluorescence PCR on the LightCycler has displayed its marked ability to detect single nucleotide substitution [1, 2]. However, many inherited diseases are caused by small deletion mutations as well as single nucleotide mutations.

In this report, we present the methods for rapid genotyping of such small deletion mutations. First, using a probe-based hybridization method, we detected a 2-bp deletion mutation in the DNA of a patient with Fabry disease, which is an X-linked recessive disorder caused by the deficient activity of α-galactosidase (α-Gal;EC 3.2.1.22). We were able to clearly genotype his family. Next, using an SYBR Green I-based method without hybridization probes, we detected a 9-bp deletion mutation in the cDNA of a patient with carbamoylphosphate synthetase I (CPS1; EC 6.3.4.16) deficiency, which is an autosomal recessive disorder affecting the first enzyme step of the urea cycle [3].

---

* Toshimitsu Niwa (✉) (e-mail: tniwa@med.nagoya-u.ac.jp)
  Department of Clinical Preventive Medicine, Nagoya University, School of Medicine,
  65 Tsuruma-Cho, Showa-ku, Nagoya 466–8550, Japan

Materials

Equipment LightCycler instrument (Roche Molecular Diagnostics, Japan)

Reagents M-MLV reverse-transcriptase (Gibco BRL)
Random primers (TAKARA, Japan)
RNase-inhibitor (Wako, Japan)
pGEM-T easy plasmid (Promega)
Amplification primers (Nihon Gene Research Laboratories Inc., Sendai, Japan)
Hybridization probes (Nihon Gene Research Laboratories Inc.)
High Pure PCR Template Purification Kit (Roche Molecular Diagnostics)
LightCycler-DNA Master SYBR Green I (Roche Molecular Diagnostics)
LightCycler-DNA Master Hybridization Probes (Roche Molecular Diagnostics)

Procedures

For the detection of a 2-bp deletion using the hybridization probe-based method, we followed the procedure described in the next section.

Sample Preparation The patient was a 15-year-old boy with classic Fabry disease who suffered from angiokeratoma, acroparesthesias, and attacks of pain in his legs. We had already identified a 2-bp deletion mutation at nucleotide 11008–11009 in his *GLA* gene (GenBank Accession #X14448) by RT-PCR and sequencing [4, 5]. This change led to a frameshift, which was described previously in another patient with the same disease [6]. We extracted genomic DNA from peripheral blood lymphocytes of the patient, his mother, his unaffected brother, and his maternal grandmother, using the High Pure PCR Template Purification Kit according to the manufacturer's instructions.

Oligonucleotides For fluorescence PCR analysis, we prepared two PCR primers and two fluorescence probes (Table 1). A 25-mer oligonucleotide probe, Anchor Probe, was labeled at the 5′ end with the LCRed640 fluorophore, which was modified at the 3′ end by phosphorylation to avoid extension. The other 20-mer oligonucleotide probe, Mutation Detection Probe, was synthesized to anneal the 2-bp deletion mutation region, which was labeled at the 3′ end with fluorescein. The distance between the two probes was one oligonucleotide. When both probes hybridize in close proximity, fluorescence resonance energy transfer occurs, producing a specific fluorescence emission of LCRed640 as a result of fluorescein excitation.

The following master mix was used for the reaction:

LightCycler PCR

	Volume [μl]	[Final]
LightCycler-DNA Master Hybridization Probes	2.0	1×
MgCl$_2$ stock solution	2.4	4 mM
Primers (10 μM)	0.4+0.4	0.2 μM
Hybridization probes (10 μM)		
Mutation detection probe	0.4	0.2 μM
Anchor probe	0.8	0.4 μM
H$_2$O (PCR grade)	11.6	
Total volume	18.0	

To complete the amplification mixtures, 18 μl of master mix and 2 μl of the genomic DNA template (30 ng) were added to each capillary.

The thermal cycling was carried out as follows:

- Denaturation at 95°C for 30 s
- Amplification

Parameter	Value		
Cycles	40		
Type	Quantification		
	Segment 1	Segment 2	Segment 3
Target temperature [°C]	95	56	72
Incubation time [s]	0	5	4
Temperature transition rate [°C/s]	20	20	20
Acquisition mode	None	Single	None
Gains	F1=1; F2=15; F3=30		

- Melting Curve Analysis

Parameter	Value		
Cycles	1		
Type	Melting curves		
	Segment 1	Segment 2	Segment 3
Target temperature [°C]	95	45	85
Incubation time [s]	30	20	0
Temperature transition rate [°C/s]	20	20	0.2
Acquisition mode	None	None	Continuous
Gains	F1=1; F2=15; F3=30		

- Cooling at 40°C for 2 min

For detection of 9-bp deletion by SYBR Green I-based method, the following procedure was followed.

Sample Preparation

We investigated a boy with CPS1 deficiency who had died at the age of 28 days of severe hyperammonemia. We checked his gene by sequencing after PCR and identified an 840G>C mutation in one allele, which led to a 9-pb deletion at nucleotide 832–840 in the cDNA as a result of aberrant splicing (GenBank Accession #Y15793). The same mutation was reported previously [7]. For this trial we prepared three templates: the RT products of the total RNAs isolated from the liver of the patient (having the 9-bp deletion heterozygously) and of a patient with ornithine transcarbamoylase (OTC) deficiency (as a control), and a plasmid (pGEM-T easy plasmid) containing the homozygous 9-bp deletion mutation. Total RNAs from the patients' livers were prepared using the modified acid guanidinium-phenol-chloroform method [8]. First-strand cDNAs were synthesized in an 8-µl mixture containing 1 µg total RNA, 240 U M-MLV reverse transcriptase, 4 µl 5 X reverse transcription buffer, 0.5 µg random primers, 0.25 mM dNTP, and 30 U RNase-inhibitor. The reaction was performed at 37°C for 1 h.

Oligonucleotides

For fluorescence PCR analysis, we prepared two PCR primers (see Table 1).

LightCycler PCR

The following master mix was used for the reaction:

	Volume [µl]	[Final]
LightCycler-DNA Master SYBR Green I	2.0	1×
MgCl$_2$ stock solution	2.4	4 mM
Primers (10 µM)	0.4+0.4	0.2 µM
H$_2$O (PCR grade)	12.8	
Total volume	18.0	

Table 1. Oligonucleotides

Detection of 2-bp deletion in the *GLA* gene (GenBank Accession #X14448)				
	Position	Length	GC (%)	T$_m$ (°C)
Primers				
GGGCCACTTATCACTAGTTGC	10928	21	52.4	70.8
TGATGAAGCAGGCAGGAT	11114R	18	50.0	64.5
Product	10928–11114	187		
Probes				
GTGGGAACGACCTCTCTCAG-F	10995	20	50.0	72.3
LCRed640-CTTAGCCTGGGCTGTAGCTATGATA-P	11016	25	48.0	74.2
Detection of 9-bp deletion in the *CPS1* gene (GenBank Accession #Y15793)				
Primers				
GCAGAACCACTAATTCAG	811	18	44.4	62.2
GCTCCTTGCGATCACTCT	865R	18	55.6	66.8
Product	811–865	55		

To complete the amplification mixtures, 18 μl of master mix and 2 μl of the templates were added to each capillary.

The thermal cycling was carried out as follows:
- Denaturation at 95°C for 30 s
- Amplification

Parameter	Value		
Cycles	40		
Type	Quantification		
	Segment 1	Segment 2	Segment 3
Target temperature [°C]	95	56	72
Incubation time [s]	0	5	2
Temperature transition rate [°C/s]	20	20	20
Acquisition mode	None	None	Single
Gains	F1=1; F2=15; F3=30		

- Melting Curve Analysis

Parameter	Value		
Cycles	1		
Type	Melting curves		
	Segment 1	Segment 2	Segment 3
Target temperature [°C]	95	60	95
Incubation time [s]	30	20	0
Temperature transition rate [°C/s]	20	20	0.2
Acquisition mode	None	None	Continuous
Gains	F1=1; F2=15; F3=30		

- Cooling at 40°C for 2 min

Results

Detection of 2-bp Deletion by Hybridization Probe-Based Method

Derivative melting curves clearly demonstrated the difference in their genotypes (Fig. 1). The curves of the unaffected brother and maternal grandmother showed a wild-type pattern with peaks at 65.5°C and 64.7°C, respectively. On the other hand, because the 2-bp internal deletion decreased the T_m, the patient with Fabry disease showed a single peak at 60.3°C, which is approximately 5°C lower than the peaks in the wild type, demonstrating that he had the hemizygous mutation of the *GLA* gene. His mother showed a heterozygous pattern with two peaks at 59.9°C and 65.7°C, demonstrating that she was a carrier of the mutation.

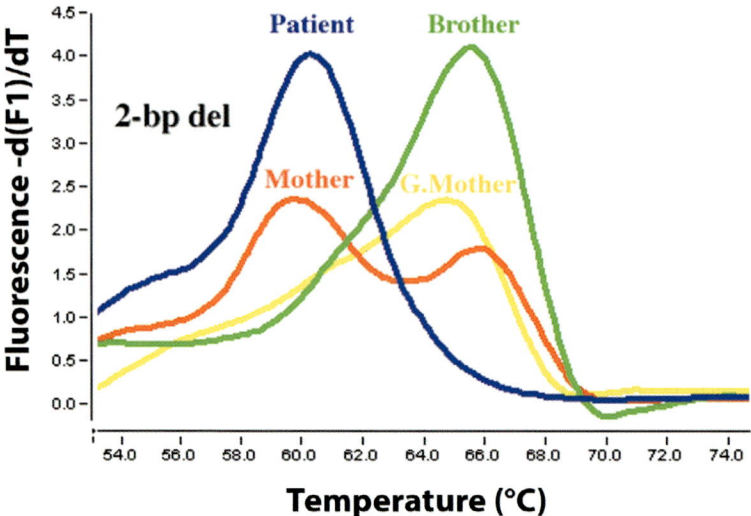

Fig. 1. The derivative melting curves for detection of 2-bp deletion in the family with Fabry disease. Patient (*blue*), mother (*red*), brother (*green*), maternal grandmother (*yellow*)

The melting temperatures were:

	Patient	Mother	Brother	Grandmother
T_m (°C)	60.3	59.9 and 65.7	65.5	64.7

Detection of 9-bp Deletion by SYBR Green I-Based Method

Derivative melting curves also clearly demonstrated their genotypes (Fig. 2). The curve of the control showed a wild-type pattern with a single peak at 76.6°C. On the other hand, the plasmid having the homozygous 9-bp deletion showed a single peak at 74.9°C, 1.8°C lower than the wild type. The patient with CPS1 deficiency showed a heterozygous pattern with two peaks at 74.9°C and 76.6°C.

The melting temperatures were:

	Patient (heterozygous)	Control	Plasmid (mutation homozygous)
T_m (°C)	74.9 and 76.6	76.6	74.9

Comments

In this report, we demonstrate that deletions of small nucleotides could be detected rapidly and easily by fluorophore techniques. We were able to distinguish the genotypes in a family with Fabry disease. It is possible that the mutation detection probe, which spanned the deletion region, hybridized to the mutation template,

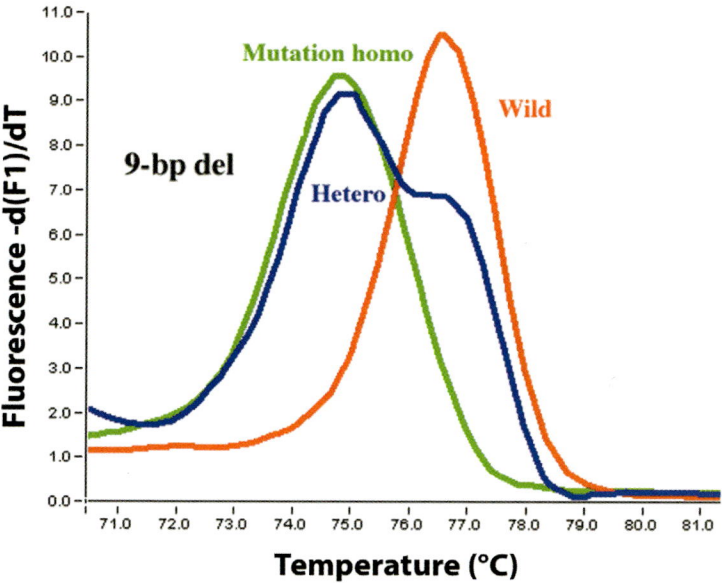

Fig. 2. The derivative melting curves for detection of 9-bp deletion. Patient (*blue*), control (*red*), plasmid-containing homozygous deletion mutation (*green*)

forming a loop with the surplus nucleotides. Therefore, it melted off from the mutation template at a lower temperature than from the wild-type template. This technique may be applied to cases with deletion mutations of larger numbers of nucleotides. The T_m of a PCR product depends on the length itself and GC content. If the size of deletions within PCR products were at least 9 bp, they could be distinguished from each other using only SYBR Green I, as shown in the CPS1 deficiency. This is a very simple procedure that does not require the design of specific hybridization probes. However, the deletion must be large enough compared to the entire fragment or contain a high GC content to influence the T_m. For example, the 9-bp deletion was detectable within a 55-bp PCR product ($T_m = 1{,}8\ °C$) but not within a 100-bp fragment ($T_m = 0{,}6\ °C$). The 9-bp product may be the minimum detectable by this method. Although the two mutations in this report are not common, we believe that these techniques can be widely used for rapid and simple screening of diseases that show common deletion mutations.

References

1. Lay MJ, Wittwer CT (1997) Real-time fluorescence genotyping of factor V Leiden during RapidCycle PCR. Clin Chem 43:2262–2267
2. Bernard PS, Lay MJ, Wittwer CT (1998) Integrated amplification and detection of the C677T point mutation in the methylenetetrahydrofolate reductase gene by fluorescence resonance energy transfer and probe melting curves. Anal Biochem 255:101–107

3. McReynolds JW, Crowley B, Mahoney MJ, Rosenberg LE (1981) Autosomal recessive inheritance of human mitochondrial carbamyl phosphate synthetase deficiency. Am J Hum Genet 33:345–353
4. Miyazaki T, Kajita M, Ohmori S, Mizutani N, Niwa T, Murata Y, et al. (1998) A novel mutation (E358K) in the ?-galactosidase A gene detected in a Japanese family with Fabry disease. Human Mutat [Suppl 1]:S139–S140
5. Aoshima T, Sekido Y, Miyazaki T, Kajita M, Mimura S, Watanabe K, Shimokata K, Niwa T. Rapid detection of deletion mutations of inherited metabolic diseases by melting curve analysis with LightCycler. Clin Chem 2000 Jan;46(1):119–22.
6. Germain D, Biasotto M, Tosi M, Meo T, Kahn A, Poenaru L (1996) Fluorescence-assisted mismatch analysis (FAMA) for exhaustive screening of the alpha-galactosidase A gene and detection of carriers in Fabry disease. Hum Genet 98:719–726
7. Hoshide R, Matsuura T, Haraguchi Y, Endo F, Yoshinaga M, Matsuda I (1993) Carbamyl phosphate synthetase I deficiency. One base substitution in an exon of the CPS I gene causes a 9-basepair deletion due to aberrant splicing. J Clin Invest 91:1884–1887
8. Chomczynski P, Sacchi N (1987) Single-step method of RNA isolation by acid guanidinium thiocyanate-phenol-chloroform extraction. Anal Biochem 162:156–159

Genotyping of the Methionine-Valine Polymorphism at Codon 129 of the Human Prion Protein by Melting Point Analysis of Fluorescently Labeled Hybridization Probes

Siegfried Burggraf*, Siegfried Kösel, Sabine Lohmann, Reinhard Beck, Bernhard Olgemöller

Introduction

Human prion diseases include Creutzfeldt-Jakob disease (CJD), the Gerstmann-Sträussler-Scheinker syndrome, fatal familial insomnia, and kuru. Since 1996 a new variant of CJD (vCJD) has been known, which is associated with the bovine spongiform encephalopathy (BSE). Prion diseases are characterized by the accumulation of abnormal prion protein deposits within the brain. The inherited forms are caused by mutations in the human prion protein gene on chromosome 20. In addition, a common methionine/valine polymorphism at codon 129 of the prion protein gene modulates disease susceptibility and phenotypic variability of human prion diseases [1–4]. Several studies have shown that methionine homozygocity at codon 129 is significantly more frequent in sporadic and iatrogenic CJD compared with the genotype distribution of the Caucasian population and confers an increased risk of developing CJD [1, 5–7]. Interestingly, all vCJD patients tested were methionine homozygotes at codon 129 [8, 9].

Here we present an assay for fast determination of the human prion protein genotype at codon position 129 by the use of rapid cycle PCR and melting point analysis of fluorogenic hybridization probes. During evaluation of the assay we encountered an artifact that can lead to incorrect results. We describe troubleshooting experiments and asymmetric PCR as a way to obtain reliable genotyping results.

Materials

LightCycler instrument — Equipment

QIAamp DNA Blood Mini Kit (Qiagen, Hilden, Germany) — Reagents
Amplification Primer (TIB MOLBIOL, Berlin, Germany)
Hybridization Probes (TIB MOLBIOL)
LightCycler-FastStart DNA Master Hybridization Probes (Roche Diagnostics, Mannheim, Germany)

---

* Siegfried Burggraf (✉) (e-mail: burggraf@labor-bo.de)
  Labor Becker, Olgemöller und Kollegen, Führichstr. 70, 81671 München, Germany

Procedure

Sample Preparation

Genomic DNA was extracted from either 200-µl samples of blood containing anticoagulant (EDTA blood) or cotton swab samples of mouth mucosa with the QIAamp DNA Blood Mini Kit, according to the manufacturer's instructions. Cotton swabs were eluted in 1 ml 0.1 M Tris/HCl, pH 8.3; 1% Tween 20. We used 200 µl of the eluant for DNA extraction. DNA was eluted from the columns with 50 µl H_2O. The concentration of the DNA was not determined. For PCR, DNA from EDTA blood was diluted 1:5 with H_2O; swab DNA was used without further dilution.

Primer and Probe Design

Primers and probes were designed using a beta version of the LightCycler probe design software (version 0.99.11; Roche Molecular Biochemicals, Mannheim, Germany). This novel software tool screens the target sequence for suitable primers and probes and calculates their melting temperatures (T_m). For the probe complementary to the region with the variable nucleotide, the software provides the T_m for match and mismatch. Two different pairs of primers and hybridization probes were designed. The sequences of all oligonucleotides and their binding sites on the human prion protein gene (GenBank accession #HSU29185) are given in Table 1. Figure 1 shows the binding sites of the two different sets of hybridization probes. Both probe systems bind to the target with a gap of two bases between anchor and detection probes. All oligonucleotides were purchased from TIB MOLBIOL.

Artificial Complements for Evaluation of Probe Quality

Two different oligonucleotides (complement–methionine and complement–valine; Table 1) were designed to be complementary to hybridization probe system 1 (anchor 1 and detection probe 1; Table 1; Fig. 1). Each complement contained seven additional bases at the 5′-end and six additional bases at the 3′-end of the probe binding region. The complements were synthesized and HPLC-purified by TIB MOLBIOL.

Restriction Fragment Length Polymorphism Analysis

The genotype of different samples was verified by restriction fragment length polymorphism (RFLP) analysis of products from a conventional PCR. It was not possible to analyze PCR products from the LightCycler reaction, since the LightCycler reaction mix contains dU instead of dT and dU is not recognized by the restriction enzyme used for RFLP.

DNA for RFLP analysis was amplified by conventional PCR in a Perkin Elmer model PE9600 thermocycler. The 50-µl PCR reaction mix contained 20 pmol each of forward primer and reverse primer (primer set 1), 0.2 mM of each dNTP, 3 mM $MgCl_2$, 5 µl PCR buffer (10x, Qiagen), 0.25 µl HotStarTaq DNA polymerase (Qiagen) and 5 µl genomic DNA. The PCR thermocycling profile included 15 min at 95°C, followed by 35 cycles of 94°C for 30 s, 55°C for 30 s and 72°C for 30 s. We added 1 µl of restriction buffer M and 0.5 µl of restriction endonuclease NspI (both Roche Molecular Biochemicals) to 10 µl PCR product and incubated at 37°C for 1 h. Fragments were separated on 3% agarose gels and visualized under UV light after staining with ethidium bromide. The following fragment pattern was obtained for the different genotypes: 169 bp and 16 bp for valine homozygous samples; 94 bp, 75 bp and 16 bp for methionine homozygous samples; and 169 bp, 94 bp, 75 bp and 24 bp for methionine/valine heterozygous samples.

Table 1. Primer sequences and hybridization probes for detection of human prion protein genotype at codon position 129

Human prion protein gene (GenBank Accession #HSU29185)				
	Position	Length	GC (%)	T_m (°C)
Primer set 1				
GTGGAACAAGCCGAGT	25743	16	56.2	59.60
GGTTGGGGTAACGGTG	25927R	16	62.5	60.53
Product	25743–25927	185		
Probe set 1				
CCGAAATGTATGATGGGCCTGCTCAT-F	25874R	26	50.0	69.25
LCRed640 CACTTCCCAGCACGTAGCC-P	25846R	19	63.2	64.82
Primer set 2				
CACAGTCAGTGGAACAAG	25735	18	50.0	58.75
GTACACTTGGTTGGGGT	25935R	17	52.9	59.13
Product	25735–25935R	201		
Probe set 2				
AGCACGTAGCCGCCA-F	25838	15	66.7	63.98
LCRed640-GCCCCCCACCACTGCCC-P	25821	17	82.4	70.40
Complement–valine				
CCTTGGCGGCTACGTGCTGGGAAGTGCCAT-GAGCAGGCCCATCATACATTTCGGCAGTGA	25821	60	58.3	
Complement–methionine				
CCTTGGCGGCTACATGCTGGGAAGTGCCAT-GAGCAGGCCCATCATACATTTCGGCAGTGA	25821	60	56.7	

Variable nucleotides in the detection probes are underlined.

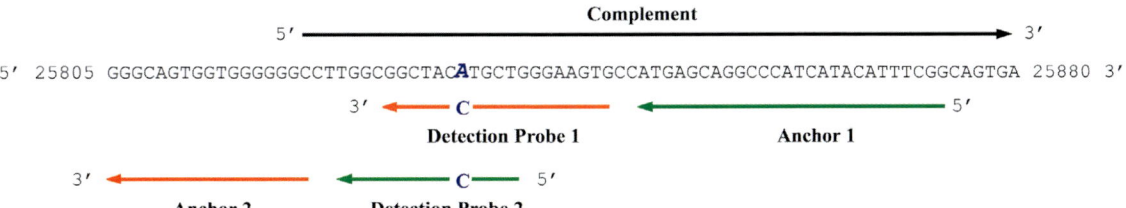

Fig. 1. Binding sites for hybridization probes and artificial complement used in the study. Numbering refers to the sequence of the human prion protein gene (GenBank accession #HSU29185; methionine at codon 129). The polymorphic site is shown in *boldface*. Probes labeled with fluorescein and LCRed640 are shown in *green* and *red*, respectively. The detection probes are complementary to the sequence coding for valine at codon 129 of the prion protein and therefore contain a cytosine at the polymorphic site

LightCycler PCR

The following mix was used for melting point analysis with artificial complements:

	Volume [µl]	Final
LightCycler FastStart DNA Master Hybridization Probes	2	1×
MgCl$_2$ (25 mM)	1.6	3.0 mM
Probes (10 µM)	0.4+0.4	0.2 µM
H$_2$O (PCR grade)	14.6	
Complement (10 µM)	1	0.5 µM
Total volume	20	

The following master mix was used to determine the methionine/valine polymorphism at position 129 of the human prion protein with hybridization probes:

	Volume [µl]	Final
LightCycler FastStart DNA Master Hybridization Probes	1	1×
MgCl$_2$ (25 mM)	0.8	3 mM
Forward primer (10 µM)	0.5	0.5 µM
Reverse primer (0.25 µM, 0.5 µM, 1.25 µM, 2.5 µM)	2	0.05–0.5 µM
Probes (10 µM)	0.2+0.2	0.2 µM
H$_2$O (PCR grade)	2.8	
Total volume	7.5	

A total of 7.5 µl of master mix and 2.5 µl of DNA (concentration not determined) were added to each glass capillary placed in precooled adaptors. Sealed capillaries were centrifuged with the adaptors (1000× *g* for 1 min) and placed into the LightCycler rotor.

The following PCR protocol was used for amplification:
- Denaturation of DNA and activation of FastStart polymerase for 10 min at 95°C
- Amplification

Parameter	Value		
Cycles	45		
Type	Quantification		
	Segment 1	Segment 2	Segment 3
Target temperature [°C]	95	55	72
Incubation time [s]	0	5	10
Temperature transition rate [°C/s]	20	20	20
Acquisition mode	None	Single	None

- Melting Curve Analysis

Parameter	Value		
Cycles	1		
Type	Melting curve		
	Segment 1	Segment 2	Segment 3
Target temperature [°C]	95	45	75
Incubation time [s]	10	60	0
Temperature transition rate [°C/s]	20	20	0.1
Acquisition mode	None	None	Cont.

Results

Using a LightCycler high-speed PCR protocol for amplification of different DNA samples with equimolar concentrations of primer set 1 and probe set 1, we detected a strong fluorescence signal at the end of the PCR amplification. Agarose gel electrophoresis of the products showed unique bands with the expected size of 185 bp (data not shown). However, the results of the melting point analysis were unexpected. Both methionine/valine heterozygous samples and a valine homozygous sample lacked a valine peak, which was supposed to occur at approximately 65°C (i.e., the calculated melting temperature for the detection probe without mismatch).

Use of Artifical Complements for Evaluation of Probe Quality

To evaluate the quality of the hybridization probes, we used single-stranded oligonucleotides complementary to probe set 1 in a melting point experiment without prior PCR amplification. Both so-called complements, one corresponding to the valine genotype (no mismatch) and the other corresponding to the methionine genotype (one mismatch), showed the calculated melting peaks of approximately 65°C and 58°C, respectively (Fig. 2).

Asymmetric PCR to Produce Single-Stranded Template for Melting Point Analysis

Since the probes had now been shown to work well on a single-stranded template, an asymmetric PCR system was devised to increase the amount of single-stranded PCR product available as target for the probes. In this system, the target DNA was amplified in the presence of a fixed amount of forward primer and various dilutions of reverse primer. A heterozygous (methionine/valine) sample could be successfully amplified with all concentrations of the reverse primer, from 5 pmol (equal to forward primer) to only 0.5 pmol (Fig. 3a, c).

Melting point analysis of the PCR reaction containing equal amounts of primers revealed only one peak around 58°C (Fig. 3b, reaction 2). The correct result for the heterozygous sample, with the two expected peaks (approximately at 58°C and 68°C), was obtained only with the lower concentration of the reverse primer (Fig. 3b, reactions 3–5).

Similar PCR experiments were performed with the methionine homozygous (Fig. 4) and the valine homozygous sample (Fig. 5). Melting point analysis revealed the expected peak around 58°C in all reactions with the methionine homozygous

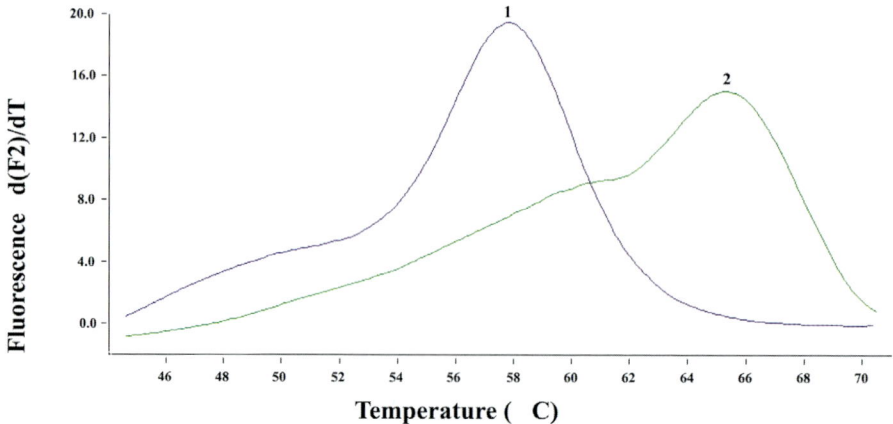

Fig. 2. Melting point analysis of artificial complements using hybridization probe set 1. The *numbered curves* show results obtained with: 1 (complement with one mismatch corresponding to the methionine genotype) and 2 (complement without mismatch corresponding to the valine genotype). Melting peaks around 58°C and 65°C; $-d(F2)/dT$, negative derivative of fluorescence with respect to temperature

sample (Fig. 4b). However, for the valine homozygous sample, only the asymmetric PCR conditions produced the correct result with a single melting peak of around 67°C (Fig. 5b, reactions 2–4).

Influence of the Amount of PCR Product on the Result of the Melting Point Analysis

After PCR amplification of a heterozygous sample, the PCR product was removed from the capillary and a dilution series of the product was performed using a PCR master mix without DNA as diluent. The diluted PCR product was filled in fresh capillaries and a melting point analysis was performed (Fig. 6). The correct genotype was only obtained with a 1:10 dilution of the PCR product.

Influence of Different Primers and Probes on the Result of the Melting Point Analysis

In order to check the influence of the probe composition, a second set of hybridization probes (probe set 2) was designed (Table 1; Fig. 1). This alternative probe system produced identical results in the melting point analysis of the heterozygous sample, i.e., only asymmetric PCR revealed the correct genotype (data not shown). The results were also reproduced when a different primer set (Table 1; primer set 2) was combined with hybridization probe system 1 (data not shown).

Assay Validation on Different Samples

In order to prove the reliability of the assay, nine different samples were genotyped with the first set of probes and asymmetric amplification conditions (Fig. 7). The result was identical to that of an RFLP analysis.

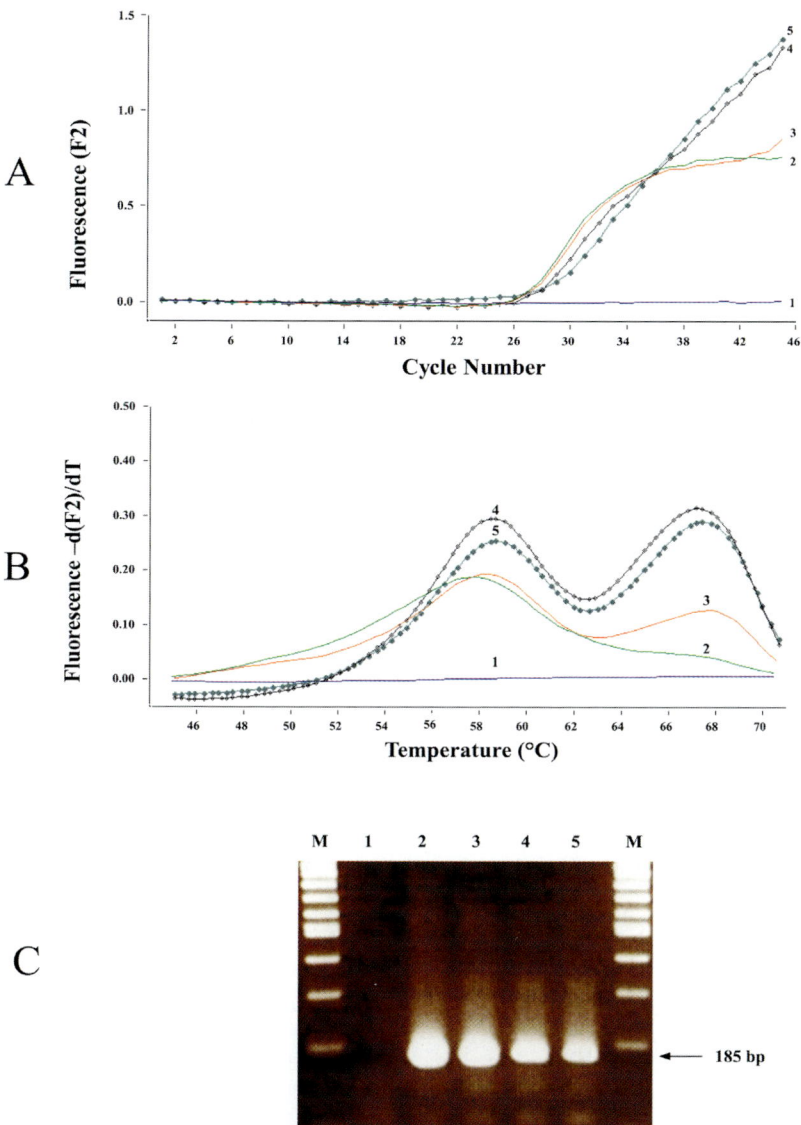

Fig. 3A–C. Rapid cycle PCR and genotyping by melting point analysis of a heterozygous sample using probe set 1 and different amounts of reverse primer 1. Each amplification contained 5 pmol of forward primer. The *numbered curves* and *lanes* show results obtained in the presence of these amounts of reverse primer: 1 (5 pmol; negative control without DNA); 2 (5 pmol); 3 (2.5 pmol); 4 (1 pmol); 5 (0.5 pmol). A Fluorescence signal detected at the annealing step of each cycle during the amplification. B Melting point analysis of the PCR products. Melting peaks between 57°C and 59°C and approximately 67–68°C. C Agarose gel electrophoresis of PCR products removed from the glass capillaries. *M*, 100-bp molecular weight marker; $-d(F2)/dT$, negative derivative of fluorescence with respect to temperature

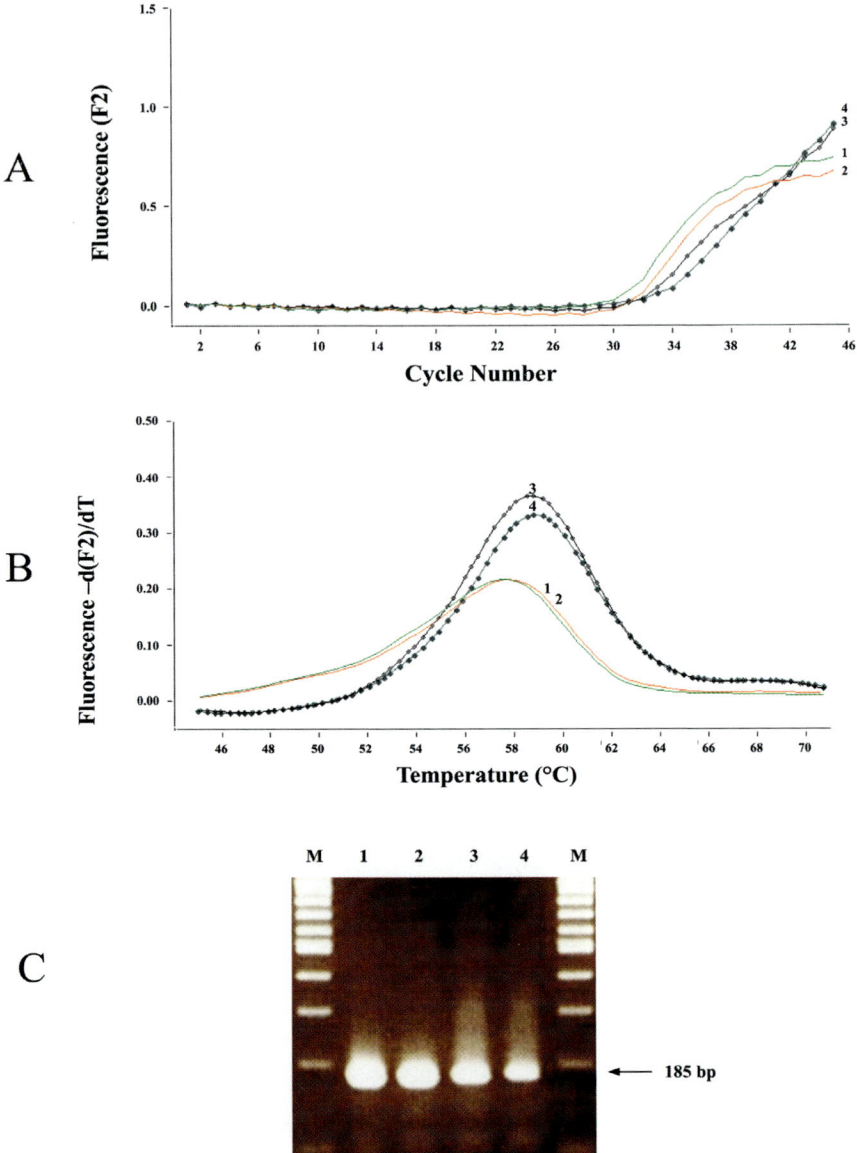

Fig. 4A–C. Rapid cycle PCR and genotyping by melting point analysis of a methionine homozygous sample using probe set 1 and different amounts of reverse primer 1. Amplification contained 5 pmol of forward primer. The numbered curves and lanes show results obtained in the presence of these amounts of reverse primer: 1 (5 pmol); 2 (2.5 pmol); 3 (1 pmol); 4 (0.5 pmol). **A** Fluorescence signal detected at the annealing step of each cycle during the amplification. **B** Melting point analysis of the PCR products. Melting peaks between 57°C and 59°C. **C** Agarose gel electrophoresis of PCR products removed from the glass capillaries. *M*, 100-bp molecular weight marker; *–d(F2)/dT*, negative derivative of fluorescence with respect to temperature

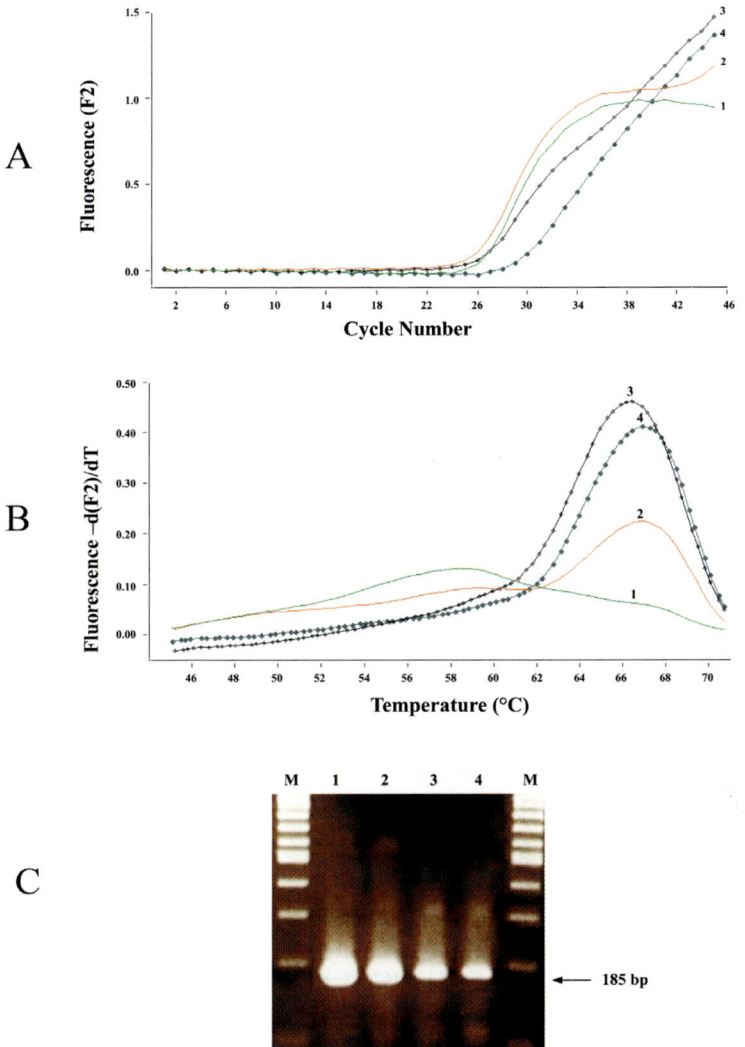

Fig. 5. Rapid cycle PCR and genotyping by melting point analysis of a valine homozygous sample using probe set 1 and different amounts of reverse primer 1. Amplification contained 5 pmol of forward primer. The *numbered curves and lanes* show results obtained in the presence of these amounts of reverse primer: 1 (5 pmol); 2 (2.5 pmol); 3 (1 pmol); 4 (0.5 pmol). **A** Fluorescence signal detected at the annealing step of each cycle during the amplification. **B** Melting point analysis of the PCR products. Melting peak approximately 67°C. **C** Agarose gel electrophoresis of PCR products removed from the glass capillaries. *M*, 100-bp molecular weight marker; $-d(F2)/dT$, negative derivative of fluorescence with respect to temperature

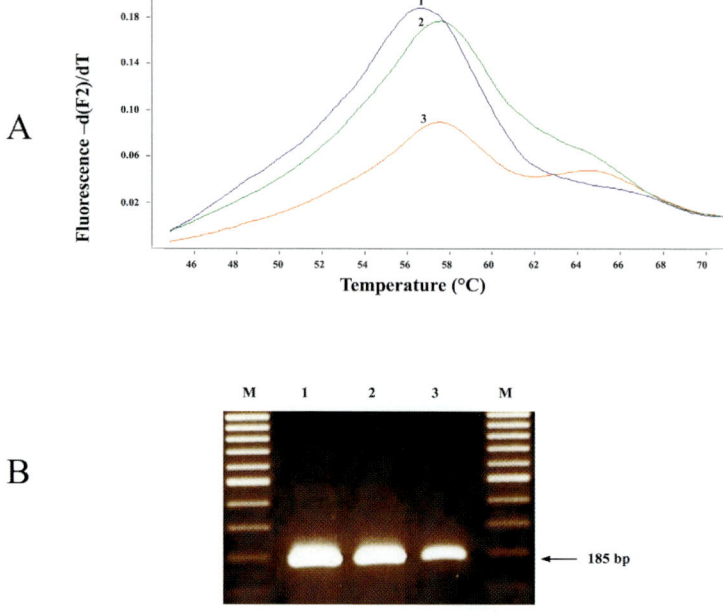

Fig. 6. Genotyping by melting point analysis of a dilution series of a PCR product from a heterozygous sample (primer/probe set 1; both primers 5 pmol). The assignment of the different samples to the curves and lanes is shown by *numbers*: 1 (undiluted PCR product); 2 (1:2 dilution of the PCR product); 3 (1:10 dilution of the PCR product). **A** Melting point analysis. Melting peaks approximately 57–58°C and 65°C. **B** Agarose gel electrophoresis of PCR products removed from the glass capillaries. *M*, 100-bp molecular weight marker; *–d(F2)/dT*, negative derivative of fluorescence vs temperature

Comments

For rapid determination of the methionine/valine polymorphism at codon 129 of the human prion protein, we developed a LightCycler rapid cycle PCR assay using melting point analysis of fluorescently labeled hybridization probes. Probes were designed with the help of a novel software tool and therefore all recommendations for proper design of mutant-detecting hybridization probes [10, 11] were followed. However, when we used a PCR protocol that was basically identical to many other published LightCycler genotyping protocols (with equimolar amounts of primers), we obtained incorrect genotyping results for heterozygous and valine homozygous samples.

A test with artificial complements proved the quality and correct design of the probes (Fig. 2). PCR amplification was excellent for all samples, as shown by the exponential increase of the fluorescence signal during the PCR (e.g., Fig. 3a) and by strong single bands on agarose gels (e.g., Fig. 3c).

Under asymmetric PCR conditions an exponential increase of the fluorescence signal could be observed until the last cycle of the PCR (e.g., Fig. 3a, reactions 4

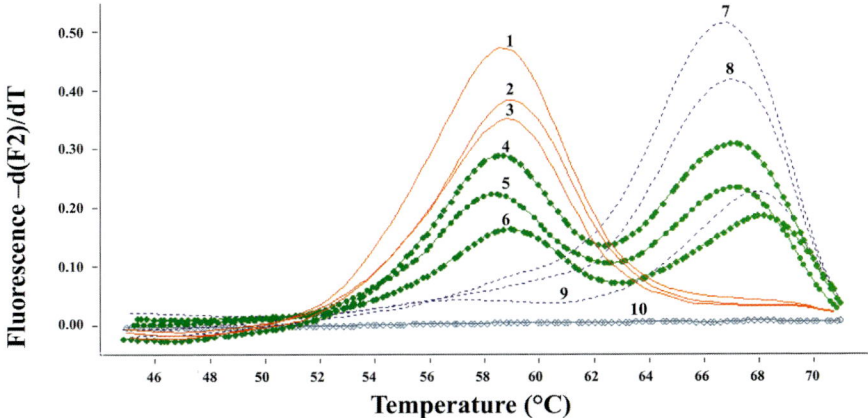

Fig. 7. Genotyping of different samples by asymmetric rapid cycle PCR and melting point analysis. Amplification contained 5 pmol of forward primer and 1 pmol reverse primer (primer set 1). The *numbered curves* show the results obtained with the following samples: 1–3 (methionine homozygous samples); 4–6 (heterozygous samples); 7–9 (valine homozygous sample); 10 (H_2O). Hybridization probe set 1 was used for the melting point analysis. Melting peaks approximately 59°C and 67–68°C. $-d(F2)/dT$, negative derivative of fluorescence with respect to temperature

and 5). The fluorescence signal of the symmetric PCR reactions reached a plateau (e.g., Fig 3a, reactions 2 and 3) or even showed a slight decrease, the so-called hook effect [12] (Fig. 5a, reaction 1), about ten cycles earlier. Therefore, the fluorescence signal at the end of the PCR was always stronger in the asymmetric PCR reactions, although the symmetric PCR reactions contained more PCR product, as shown by the stronger bands on the agarose gels. This can be explained by a reannealing of both strands of the PCR product in later PCR cycles, which competes with binding of the hybridization probes [12].

Asymmetric PCR also allowed accurate genotype analysis. These results suggest that the high GC content of the hybridization probe target region (63% G+C from position *25805* to *25874*; Fig. 1) promotes a reannealing of equimolar amounts of the PCR product strands and inhibits the binding of the detection probe to the valine genotype. However, the replacement of G with A in the methionine genotype allowed probe hybridization. The net result was a single methionine melting peak for a heterozygous sample (Fig. 3b, reaction 2) and no melting peak at the calculated temperature of 65°C in a valine homozygous sample (Fig. 5b, reaction 1).

Using a different hybridization probe system and a different primer set (probe set 2 and primer set 2; Table 1), we showed that our results were not a consequence of incorrect primer or probe design. The alternative primer and probe set gave exactly the same results, i.e., correct genotyping of the samples only with asymmetric PCR.

As an alternative to asymmetric PCR, it is also possible to get correct results by decreasing the amount of PCR product in order to favor probe binding over strand reannealing, as demonstrated in a dilution experiment of a PCR product (Fig. 6).

However, it is not easy to develop a PCR protocol to get a similar low amount of PCR product for all samples, especially if one wants to avoid determining the template DNA concentration for every sample. Therefore the asymmetric PCR approach seems to be the most practicable.

In summary, our optimized PCR protocol (asymmetric PCR with more forward than reverse primer) allows fast and specific genotyping of the human prion protein polymorphism at codon 129 in research samples (Fig. 7).

The transmissible new variant of CJD has been known for only a few years and the number of patients is still low. Further studies are necessary to evaluate whether vCJD occurs in methionine homozygous individuals only, as incubation periods may be longer in methionine/valine heterozygotes and valine homozygotes [8, 13]. In this context our LightCycler assay for rapid genotyping, especially when performed on samples obtained from the cheek mucosa (which avoids blood drawing by a physician), can provide necessary data. In addition it might help to predict an increased individual risk of developing transmissible forms of Creutzfeldt-Jakob disease.

However, our findings emphasize the need for extensive validation of new LightCycler tests. In addition, it is absolutely necessary to include controls for every possible genotype in each run. With this in mind, a melting point analysis using the LightCycler is an excellent alternative to other more laborious and time-consuming techniques for the detection of single nucleotide polymorphisms in research applications.

References

1. Windl O, Dempster M, Estibeiro JP, Lathe R, de Silva R, Esmonde T, Will R, Springbett A, Campbell TA, Sidle KC, Palmer MS, Collinge J (1996) Genetic basis of Creutzfeldt-Jakob disease in the United Kingdom: a systematic analysis of predisposing mutations and allelic variation in the PRNP gene. Hum Genet 98:259–264
2. Parchi P, Giese A, Capellari S, Brown P, Schulz-Schaeffer W, Windl O, Zerr I, Budka H, Kopp N, Piccardo P, Poser S, Rojiani A, Streichemberger N, Julien J, Vital C, Ghetti B, Gambetti P, Kretzschmar H (1999) Classification of sporadic Creutzfeldt-Jakob disease based on molecular and phenotypic analysis of 300 subjects. Ann Neurol 46:224–233
3. Masullo C, Macchi G (2001) Does PRNP gene control the clinical and pathological phenotype of human spongiform transmissible encephalopathies? Clin Neuropathol 20:19–25
4. Hauw JJ, Sazdovitch V, Laplanche JL, Peoc'h K, Kopp N, Kemeny J, Privat N, Delasnerie-Laupretre N, Brandel JP, Deslys JP, Dormont D, Alperovitch A (2000) Neuropathologic variants of sporadic Creutzfeldt-Jakob disease and codon 129 of PrP gene. Neurology 54:1641–1646
5. Laplanche JL, Delasnerie-Laupretre N, Brandel JP, Chatelain J, Beaudry P, Alperovitch A, Launay JM (1994) Molecular genetics of prion diseases in France. French Research Group on Epidemiology of Human Spongiform Encephalopathies. Neurology 44:2347–2451
6. Deslys JP, Marce D, Dormont D (1994) Similar genetic susceptibility in iatrogenic and sporadic Creutzfeldt-Jakob disease. J Gen Virol 75:23–27
7. Salvatore M, Genuardi M, Petraroli R, Masullo C, D'Alessandro M, Pocchiari M (1994) Polymorphisms of the prion protein gene in Italian patients with Creutzfeldt-Jakob disease. Hum Genet 94:375–379
8. Ironside JW, Head MW, Bell JE, McCardle L, Will RG (2000) Laboratory diagnosis of variant Creutzfeldt-Jakob disease. Histopathology 37:1–9

9. Will RG, Zeidler M, Stewart GE, Macleod MA, Ironside JW, Cousens SN, Mackenzie J, Estibeiro K, Green AJ, Knight RS (2000) Diagnosis of new variant Creutzfeldt-Jakob disease. Ann Neurol 47:575–882
10. Bernard PS, Reiser A, Pritham GH (2001) Mutation detection by fluorescent hybridization probe melting curves. In: Meuer S, Wittwer C, Nakagawara K (eds) Rapid cycle real-time PCR: methods and applications. Springer Verlag, Berlin, Heidelberg, New York, pp 11–19
11. Landt O (2001) Selection of hybridization probes for real-time quantification and genetic analysis. In: Meuer S, Wittwer C, Nakagawara K (eds) Rapid cycle real-time PCR: methods and applications. Springer Verlag, Berlin, Heidelberg, New York, pp 35–41
12. Roche Molecular Biochemicals (1999) Decrease of fluorescent signal in the plateau phase of a LightCycler PCR – hook effect. Technical Note No. LC 8/99
13. Zeidler M, Ironside JW (2000) The new variant of Creutzfeldt-Jakob disease. Rev Sci Tech 19:98–120

Rapid Detection of Missense Mutations in the Prostatic Steroid 5α-Reductase Gene Using Real-Time Fluorescence PCR and Melting Curve Analysis

Markus Nauck*, Winfried März, Heinrich Wieland

Introduction

Prostate cancer is a very common disease in more developed countries, but its cause is largely unknown. It is an androgen-dependent cancer, and it has been proposed that variations in androgen metabolism may affect a man's risk of this disease [1]. There is accumulating evidence that increased intraprostatic androgen metabolism, particularly through the enzyme steroid 5α-reductase, may have an important role in predisposition to prostate cancer [2]. This enzyme catalyses the conversion of testosterone to dihydrotestosterone – the most potent androgen in the prostate. Thus, genetic variants encoded by the steroid 5α-reductase gene (*SRD5A2*) may affect the risk of developing prostate cancer. Makridakis et al. recently reported a missense mutation in *SRD5A2*, which results in the replacement of an alanine residue at codon 49 with threonine (A49T) [3]. This A49T amino acid substitution increased the risk of clinically significant disease 7.2-fold in African-American men and 3.6-fold in Hispanic men. This clinical association is further supported by data obtained in vitro showing that the mutant enzyme had a higher enzymatic activity, indicating an increased metabolic activation of testosterone to dihydrotestosterone in carriers of the *49T* allele. Furthermore, finasteride, a competitive inhibitor of *SRD5A2*, showed a much lower efficacy in inhibiting the enzymatic activity.

A second polymorphism in *SRD5A2*, which substitutes leucine for valine at codon 89, has also been implicated in the predisposition to prostate cancer and early tumor progression. Although the V89L substitution has been shown to result in a significantly reduced steroid 5α-reductase enzyme activity in vivo, to date no consistent differences in the frequency of the *V89L SRD5A2* gene polymorphism between cases and controls have been observed [4]. There are ongoing studies to determine the association between SRD5A2 genotypes and prostate cancer in Caucasians.

The identification of genetic variants in the *SRD5A2* gene has important implications for identification of men at risk before symptoms arise and for development of chemopreventive strategies.

---

* Markus Nauck (✉) (e-mail: msnauck@med1.ukl.uni-freiburg.de)
  University Hospital Freiburg, Department of Clinical Chemistry, Hugstetter Strasse 55, 79106 Freiburg, Germany

We describe here a homogenous assay for the simultaneous genotyping of both mutations in the *SRD5A2* gene that combines RapidCycle PCR with allele-specific fluorescent probe melting profiles on the LightCycler. In order to allow the simultaneous analysis of the two polymorphic codons in a single reaction, two reporter dyes with different excitation and emission spectra, LightCycler Red 640 (LCRed640) and LightCycler Red 705 (LCRed750), and the LightCycler color compensation software that corrects for the temperature-dependent crosstalk between the emission spectra of these dyes were utilized. The method is fast and robust and thus is ideally applicable to *SRD5A2* genotyping in both clinical and research settings.

Materials

Equipment
LightCycler instrument
Reagent amplification primer (MWG Biotech, Eberswalde, Germany)
Hybridization probes (TIB MOLBIOL, Berlin, Germany)

Reagents
The following reagents were purchased from Roche Diagnostics, Mannheim:
High-Pure PCR Template Preparation Kit
LightCycler-DNA Master Hybridization Probes
LightCycler glass capillary cuvettes

Procedure

Sample Preparation
Genomic DNA was isolated from whole blood or buffy coats using the High-Pure PCR Template Preparation Kit. The DNA was resuspended in 10 mmol/l Tris (pH 7.4), containing 0.1 mmol/l EDTA at a concentration of 20 ng/µl.

Design of Primers and Fluorogenic Probes
The amplification primers were designed using the primer3 input program from the Whitehead Institute for Biomedical Research (website: www-genome.wi.mit.edu/cgi-bin/primer/primer3.cgi). The amplified fragment harboring both polymorphic sites in exon 1 of the *SRD5A2* gene is 198 bp in size (GenBank locus L03843) (Table 1). The primers were synthesized by standard phosphoramidite chemistry (MWG-Biotech, Eberswalde, Germany). Primers and probes for the detection of the alanine-threonine substitution at codon 49 and of the valine-leucine substitution at codon 89 in exon 1 of the human *SRD5A2* gene are given in Table 1.

Two pairs of hybridization probes labeled with different fluorophores were used in one reaction in order to genotype both polymorphic codons simultaneously. For codon 49 the LCRed640-labeled detection probe was designed to be complementary to the sense strand of the *49A* allele with the polymorphic nucleotide seven bases from the 3′ end (underlined nt.), as shown in Table 1 and Fig. 1. The resulting A–C mismatch between detection probe and the *49T* allele is the most unstable mismatch possible with this polymorphism, ensuring a maximum difference in the melting temperatures (T_m) between both alleles. Due to an extremely high GC

Table 1. Oligonucleotides

Human *SRD5A2* gene (GenBank Accession #L0483)				
	Position	Length	GC (%)	T_m (°C)
Primers				
AGCACACGGAGAGCCTGAAG	847	20	60.0	64.0
CCGCAAGGGAAAAACGCTAC (R)	1044	20	55.0	61.9
Product	847–1044	198	67.7	
Probes for codon 49				
LCRED640-GCGCGGGCTGGCAGG-PH (R)	881	15	86.7	67.9
GCAGCTCCTGCAGGAACCAGGC-FL (R)	897	22	68.2	70.3
Probes for codon 89				
LCRED705-CCTCTTCTGCCTACATTACTTC-PH	998	22	45.5	58.1
CACCTGGGACGGTACTTCTGG-FL	976	21	61.9	64.5

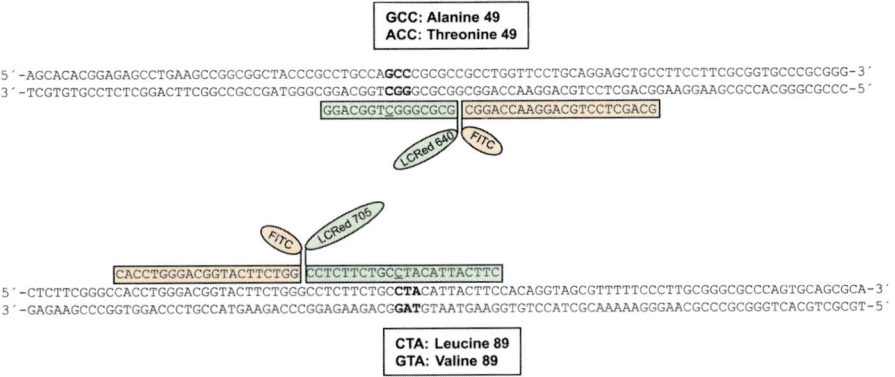

Fig. 1. Relative orientation of the fluorophore-labeled anchor and detection probes. The *A49T* polymorphism is a result of a G to A substitution at nucleotide 888 of the human *SRD5A2* gene. This polymorphism creates an A–C mismatch between the sense strand of the *49T* allele and the detection probe. This mismatch destabilizes the hybrid which results in a decrease in the melting temperature of the probe. In contrast, complete matching of detection probe and sense strand of the *49A* allele results in a higher melting temperature of the hybrid. The *V89L* polymorphism is a result of a G to C substitution at nucleotide 1008. This polymorphism creates a C–C mismatch between the antisense strand of the *89V* allele and the detection probe. In contrast, complete matching of detection probe and antisense strand of the *89L* allele results in higher stability of the hybrid

content of more than 90% the neighboring nucleotides of this polymorphic site provide a difficult environment for allele-discriminating hybridization probes.

The probe for detection of the *V89L* polymorphism was labeled with LCRed705 and hybridizes with perfect match to the antisense-strand of the 89L-allele. It is true that the resulting C–C mismatch between detection probe and the *89V* allele is rather stable, but the low GC content of the detection probe of 45% favors the allele-specific discrimination in spite of the relative stability of this mismatch.

Each of the 3′-fluorescein-labeled anchor probes binds with a distance of one base 5′ to their corresponding detection probes. All fluorophore-labeled probes were synthesized and purified by reverse-phase HPLC by TIB MOLBIOL, Berlin, Germany.

Experimental Protocol

The analysis on the LightCycler was performed in a reaction volume of 20 µl with 40 ng of genomic DNA and a master mix of the following composition:

	Volume [µl]	[Final]
LightCycler-DNA Master Hybridization Probes	2.0	1×
$MgCl_2$ (25 mM)	1.2	2.5 mM
Primers (3 µM each)	2.0	0.3 µM each
Probes (4 µM each)	2.0	0.4 µM each
H_2O (PCR grade)	10.8	
Total master mix volume per reaction	18.0	

After loading the samples into the glass capillary cuvettes, the DNA template was added (2 µl=40 ng) and the capillaries were sealed, briefly centrifuged, and then placed into the LightCycler rotor. For amplification the following thermocycling protocol was used:
- Denaturation at 95°C for 30 s
- Amplification

Parameter	Value		
Cycles	45		
Type	Quantification		
	Segment 1	Segment 2	Segment 3
Target temperature [°C]	95	60	72
Incubation time [s]	0	6	6
Temperature transition rate [°C/s]	20	20	3
Acquisition mode	None	Single	None
Gains	F1=1; F2=10; F3=30		

- Melting Curve Analysis

Parameter	Value		
Cycles	1		
Type	Melting curve		
	Segment 1	Segment 2	Segment 3
Target temperature [°C]	95	35	75
Incubation time [s]	30	150	0
Temperature transition rate [°C/s]	20	20	0.1
Acquisition mode	None	None	Cont.
Gains	F1=1; F2=10; F3=30		

Results

Forty-five cycles of amplification were performed with genomic DNA with different *SRD5A2* genotypes and a template-free control using the fluorescence resonance transfer detection system outlined in Fig. 1. The fluorescence signal was measured in channel F2 at the end of each annealing phase and increased as product accumulated. Under our conditions, the fluorescence signal appeared above background levels after 30 cycles. No increase in fluorescence signal was observed in the absence of template. When analyzed on an agarose gel, we found a pure amplification product of the expected size.

The process of hybridization and melting of the detection probes to the target was monitored by melting curve analysis. By plotting the negative derivative of the fluorescence signal with temperature versus temperature ($-(dF/dT)$ vs T), peaks are obtained at the respective melting temperatures (T_m).

The detection probes match perfectly with the alleles coding for alanine (sequence GCC) and leucine (sequence CTA) at codons 49 and 89, respectively. Accordingly, when examining codon 49, we observed a T_m at 62.5°C with a DNA homozygous for the sequence GCC, whereas a DNA coding for threonine (sequence ACC) led to a markedly lower T_m of 51.0°C. Heterozygous samples contain both types of targets and, thus, generated both peaks (Fig. 2a). The fluorescent signal acquired in channel 3 was used to genotype codon 89. However, due to the interference of fluorescence signals caused by LCRed640, which are also recorded in channel 3, the emission signal at 705 nm had to be corrected for the contribution of LCRed640 by utilizing the crosstalk compensation module of the LightCycler system. By doing so, the alleles coding for leucine (sequence CTA) and valine (sequence GTA) at codon 89 were clearly distinguishable, with melting peaks at 57.5°C and 49.0°C, respectively (Fig. 2b).

With different samples showing different amplification efficiencies, the derivative melting curves were highly reproducible with melting peaks differing by less than 0.8°C for the same allele, allowing easy and unambiguous assignment of genotypes to the respective melting curves (Table 2).

To evaluate the reliability of the fluorescence genotyping, 100 human DNA samples were genotyped for both *SRD5A2* polymorphisms by both the conventional allele-specific restriction fragment analysis and the homogenous fluorescence assay. The genotypes determined with both methods were in 100% concordance. The genotyping of the 100 samples on the LightCycler was completed within 3 h, while the ASRA protocol took 12 h and required several manual sample processing steps.

Comments

The homogeneous protocol presented here combines a number of advantages:
- The fluorescence assay is robust and reliable as documented by the complete concordance of 100 genotypes determined with both the conventional allele-specific restriction fragment analysis and the LightCycler protocol.

a) Genotyping of codon 49

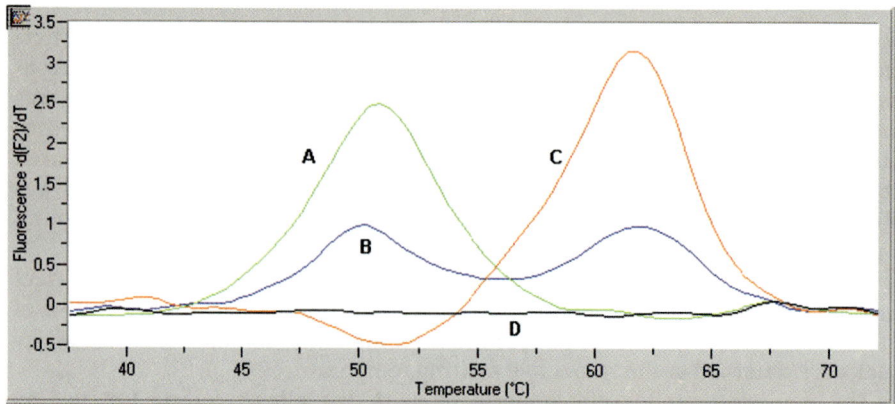

b) Genotyping of codon 89

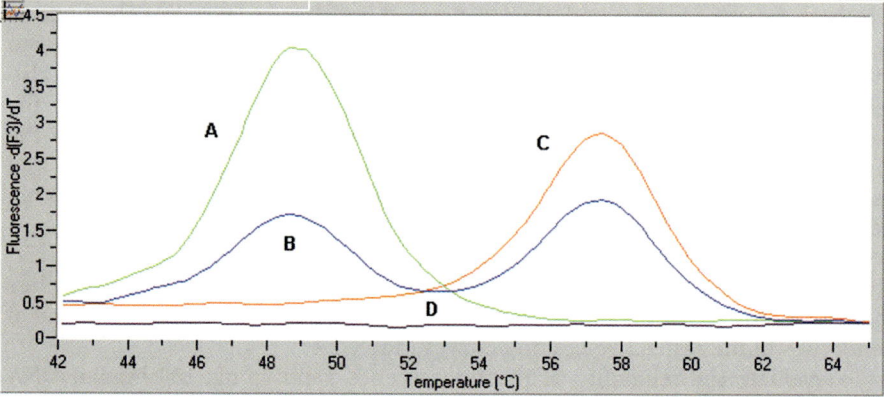

Fig. 2a,b. Genotyping of the *A49T* (**a**) and the *V89L* (**b**) polymorphisms of the *SRD5A2* gene with allele-specific fluorescent probes by derivative melting curve plots. Immediately after amplification, a melting analysis was performed. Data for these plots were obtained during the melting transition of the LightCylcer Red 640- and LightCylcer Red 705-labeled detection probes from the amplified fragment. The temperature transition was programmed at 0.1°C/s with continuous fluorescence acquisition for each sample from 35°C to 75°C. Derivative melting curve: the melting curve plot of fluorescence signal (F) vs temperature (T) was transformed into a derivative melting curve plot with dF/dT vs temperature (T). **a** Genotyping of codon 49: the melting curve is plotted for a sample homozygous for alanine (A), a heterozygous sample (B), and a sample homozygous for threonine (C). Melting analysis of a no template control (D) was also performed. **b** Genotyping of codon 89: the melting curve is plotted for a sample homozygous for leucine (A), a heterozygous sample (B), and a sample homozygous for valine (C). D shows the melting analysis of a no template control

Table 2. Melting temperatures of *SRD5A2*

Locus	Allele	Pairing	T_m (°C) observed
SRD5A2; codon 49	49 alanine	G–C match	62.5
	49 valine	A–C mismatch	51.0
SRD5A2; codon 89	89 leucine	C–G match	57.5
	89 valine	C–C mismatch	49.0

- The processing of the samples is simple and the analysis is fast, providing rapid results as well as the possibility of high throughput genotyping.
- As this method is performed in a closed system with no post-amplification processing, potential problems with sample tracking and end-product contamination are eliminated.
- As hands-on time is shorter than in any other technique used so far and costs for reagents and consumables are not higher than for conventional assays, the fluorescence method allows *SRD5A2* genotyping in a very economic manner.
- The simultaneous analysis of two polymorphisms by utilizing two different reporter dyes and the color compensation software additionally saves time and money.

In summary, this homogeneous (closed-tube) assay for rapid genotyping of the *SRD5A2* polymorphisms at codons 49 and 89 on the LightCycler is ideally applicable to routine analysis in a clinical and/or research setting. This method may facilitate further research on the association between polymorphisms in the *SRD5A2* gene and the risk of developing prostate cancer and eventually, when a predictive value has been established, it may be used in the risk stratification of patients and for selection of therapeutic strategies.

References

1. Ross RK, Pike MC, Coetzee GA, Reichardt JK, Yu MC, Feigelson H, Stanczyk FZ, Kolonel LN, Henderson BE (1998) Androgen metabolism and prostate cancer: establishing a model of genetic susceptibility. Cancer Res 58:4497–4504
2. Ross RK, Bernstein L, Lobo RA, Shimizu H, Stanczyk FZ, Pike MC, Henderson BE (1992) 5-alpha-reductase activity and risk of prostate cancer among Japanese and US white and black males. Lancet 339:887–889
3. Makridakis NM, Ross RK, Pike MC, Crocitto LE, Kolonel LN, Pearce CL, Henderson BE, Reichardt JK (1999) Association of mis-sense substitution in SRD5A2 gene with prostate cancer in African-American and Hispanic men in Los Angeles, USA. Lancet 354:975–978
4. Makridakis N, Ross RK, Pike MC, Chang L, Stanczyk FZ, Kolonel LN, Shi CY, Yu MC, Henderson BE, Reichardt JK (1997) A prevalent missense substitution that modulates activity of prostatic steroid 5alpha-reductase. Cancer Res 57:1020–1022

Applications in Oncology

**Analysis of Microsatellite Instability by Melting Peak Analysis
with BAT26 and BAT25 Specific Fluorescence Hybridization Probes** 139
Wolfgang Dietmaier, Arndt Hartmann, Ferdinand Hofstädter

**Two Color Multiplexing and Typing of Human Papillomavirus Types 16, 18
and 45 on LightCycler™** 147
Monica L. Henriquez, Brian E. Caplin, Randy P. Rasmussen

**Quantitative Analysis of AML1-ETO Fusion Transcripts
in t(8;21) Positive AML Using Real-Time RT-PCR** 159
Martin Weisser, Claudia Schoch, Torsten Haferlach,
Wolfgang Hiddemann, Susanne Schnittger

**Rapid Quantitative Detection of Free Cancer Cells
in the Peritoneal Cavity of Gastric Cancer Patients
with Real-Time CEA RT-PCR Using Hybridization Probes** 169
Hayao Nakanishi, Yasuhiro Kodera, Masae Tatematsu

**Quantitative Measurement of the mRNA Expression
of the Tumor-Associated Antigen PRAME by Real-Time RT-PCR
Using LightCycler and SYBR Green I Technology** 177
Jochen Greiner, Mark Ringhoffer, Anita Szmaragowska,
Sandra Hübsch, Hartmut Döhner, Michael Schmitt

**Expression Analysis of Telomerase-Genes hTERT
and hTR by Quantitative PCR on LightCycler** 187
Bernd Frodermann, Christopher Poremba

**Measurement of *MDR1* Gene Expression
by Real-Time Quantitative RT-PCR Using the LightCycler Instrument** 199
Chung-Che Chang, Sherrie Perkins, Carl Wittwer

Analysis of Microsatellite Instability by Melting Peak Analysis with BAT26 and BAT25 Specific Fluorescence Hybridization Probes

Wolfgang Dietmaier*, Arndt Hartmann, Ferdinand Hofstädter

Introduction

Microsatellite instability (MSI) can be detected in about 15% of all colorectal cancers (CRC) as a result of defective mismatch repair. Almost all (>90%) CRC from patients with hereditary non-polyposis colorectal cancers (HNPCC) show MSI due to mutations in the *hMSH2*, *hMLH1*, and *hMSH6* mismatch repair genes [1, 2, 3, 4]. MSI can also be observed in other tumors of the HNPCC tumor spectrum, e.g., gastric, ovarian, and endometrial carcinomas. In order to identify HNPCC patients, MSI analysis of the tumor DNA and immunohistochemical detection of mismatch repair expression in the tumor tissue is performed as a first step, followed by germline mutation analysis of the mismatch repair gene with loss of protein expression in the tumor tissue. In colorectal tumors, microsatellite analysis is performed by amplification of five microsatellite markers [5, 6], which are separated by gel or capillary electrophoresis and visualized using autoradiography [1], silver staining [7], or fluorescence techniques [8, 9]. A tumor with at least two unstable markers (2/5, 40%) is defined as MSI-H (high frequency microsatellite instability) [5, 6].

To avoid the need for a time-consuming electrophoresis step and to minimize contamination risk, we have established a fast and simple method for detection of MSI using the LightCycler. The procedure is based on a real-time amplification and subsequent melting point analysis using LCRed640- and LCRed705-labeled hybridization probes (HyProbes), which are complementary to the mononucleotide repeats of BAT26 and BAT25 markers, respectively. MSI can be displayed by shifted melting peaks due to alterations of the length of repetitive sequences.

Materials

LightCycler instrument **Equipment**

Amplification Primers (GENSET, Paris, France) **Reagents**
Hybridization Probes (GENSET)

---

* Wolfgang Dietmaier (✉) (e-mail: wolfgang.dietmaier@klinik.uni-regensburg.de)
 Molecular Pathology Diagnostic Unit, University of Regensburg, Franz-Josef-Strauß-Allee 11, 93053 Regensburg, Germany

High Pure PCR Template Preparation Kit (Roche Diagnostics, Mannheim, Germany)
LightCycler DNA Master Hybridization Probes (Roche Diagnostics)

Procedure

Sample preparation

Tumor tissue and normal mucosa from 113 patients with CRC were microdissected from 5-µm sections of formalin-fixed paraffin-embedded tissues by laser assisted (P.A.L.M., ∝Germany) or manual microdissection. DNA preparation was done using the High Pure PCR Preparation Kit (Roche Diagnostics, Mannheim, Germany) and quality of DNA was checked by electrophoresis (1.3% agarose gel). DNA concentration was spectrophotometrically measured and 50–100 ng template DNA was used for LightCycler PCR.

Oligonucleotides

Amplification of BAT26 and BAT25 markers was done with published primers (Table 1), [10]. Hybridization probes (Table 1) were designed as far as possible according to the guidelines recommended by the LightCycler operator's manual. The donor hybridization probes were labeled at the 3′ end with fluorescein

Table 1. Oligonucleotides

BAT26 GenBank Accession #U41210				
	Position	Length	GC (%)	T_m (°C)
BAT26 Primers				
TGACTACTTTTGACTTCAGCC	123	21	42.9	59.4
AACCATTCAACATTTTTAACCC	243R	22	31.8	57.1
Product 1	243–123	120		
BAT26 Probes				
GCAGCAGTCAGAGCCCTTAACCT-F	163	23	56.5	68.0
LCRed640-TCAGGTAAAAAAAAAAAAAA-AAAAAAAAAAAA-P	190	32	9.4	57.8

BAT25 GenBank Accession #L04143				
	Position	Length	GC (%)	T_m (°C)
BAT25 Primers				
TCGCCTCCAAGAATGTAAGT	6839	20	45	60.4
TCTGCATTTTAACTATGGCTC	6961R	21	38	57.5
Product 1	6739–6961	122		
BAT25 Probes				
CAAAAAAAAAAAAAAAAAAAAAAAATCA-F	6925R	29	6.9	55.0
LCRed705-AACAAAACACAAAACTCTTTAGA-GAATC-P	6892R	28	28.6	60.4

and were complementary to upstream sequences adjacent to repetitive poly adeninosine (BAT26) and poly thymidine (BAT25) stretches within the *hMSH2* and *c-kit* gene, respectively. Acceptor probes were designed to hybridize to repetitive poly thymidine sequences and to at least one unique base at the 5′ and 3′ ends of the repetitive stretches. The BAT26 and BAT25 acceptor probes were labeled with LCRed640 and LCRed705, respectively. When hybridized to the templates, donor and acceptor probes have a distance of three (BAT26) and four (BAT25) bases.

Master Mix of BAT26 and BAT25 assays were composed as follows:

LightCycler PCR

	Volume [µl]	[Final]
LightCycler DNA Master Hybridization Probes	1.5	1×
MgCl$_2$ (25 mM)	1.8	3.0 mM
Primers (25 µM each)	0.3+0.3	0.5 µM
Probes (15 µM each)	0.15+0.15	0.15 µM
H$_2$O (PCR grade)	8.8	
Total volume	13	

We combined 13 µl Master Mix of each BAT assay with 2 µl template DNA in glass capillaries in precooled adaptors. Glass capillaries were sealed and centrifuged before placing them into the LightCycler rotor.

The following LightCycler amplification protocol was used:
- Denaturation at 95°C for 90 s
- Amplification

Parameter	Value			
Cycles	50			
Type	Quantification			
	Segment 1	Segment 2	Segment 3	Segment 4
Target temperature [°C]	95	60	50	72
Incubation time [s]	0	10	3	10
Temperature transition rate [°C/s]	20	20	20	20
Acquisition mode	None	None	Single	None
Gains	F1=1; F2=10; F3=1 (BAT26)			
	F1=1; F2=1; F3=10 (BAT25)			

- Melting curve analysis BAT26

Parameter	Value		
Cycles	1		
Type	Melting curve		
	Segment 1	Segment 2	Segment 3
Target temperature [°C]	95	35	95 (BAT26)
			65 (BAT25)
Incubation time [s]	0	30	0
Temperature transition rate [°C/s]	20	20	0.2
Acquisition mode	None	None	Continuous
Gains	F1=1; F2=10; F3=1 (BAT26)		
	F1=1; F2=1; F3=10 (BAT25)		

Results

Successful amplification and melting point analysis of BAT26 was feasible in 65 of 81 (80%) tumor samples and in 64 of 81 (79%) normal mucosa samples. Melting temperature (T_m) of DNA from normal mucosa and control DNA from mononuclear blood cells was 51.0–51.5°C (Fig. 1a). In contrast, the T_m of DNA from all 16 tumors with MSI-H phenotype, as determined by fragment analysis using the ABI310 sequencer and loss of hMSH2, hMLH1 or hMSH6 protein expression, shifted to 42–50°C (Fig. 1b). Fragment analysis using the ABI310 capillary sequencer revealed that the T_m alterations reflect shortenings of 3–12 bases in the repetitive stretch of 26 adenosines within the BAT26 marker.

Amplification of BAT25 was possible in 63 of 81 (78%) tumor samples and in 60 of 81 (74%) normal mucosa samples. BAT25 showed a T_m of 47–48°C in DNA from normal mucosa, tumors without microsatellite instability, and DNA from mononuclear blood cells (Fig. 2a). In contrast, the T_m of 15 MSI-H tumors was 42.5–43.7°C (Fig. 2b), which reflects shortenings of 5–7 bases within 25 repetitive adenines of the BAT25 marker as determined by fragment analysis.

Comments

Detection of MSI in colorectal tumors can be achieved by LightCycler melting peak analysis using specific hybridization probes for BAT26 and BAT25. This new technique is significantly faster than conventional PCR techniques followed by electrophoretic separation of the amplicons. Furthermore, the contamination risk, especially in a high throughput laboratory setting, is greatly reduced. Since MSI-H is defined as the detection of at least two unstable microsatellites from a set of five markers [5, 6], it may not be necessary to analyze more markers if MSI is detected by BAT26 and BAT25, which are described as the most sensitive markers for MSI detection in colorectal tumors [5, 11, 12, 13]. Most colorectal MSI-H tumors

Analysis of Microsatellite Instability by Melting Peak Analysis with BAT26 and BAT25 Specific

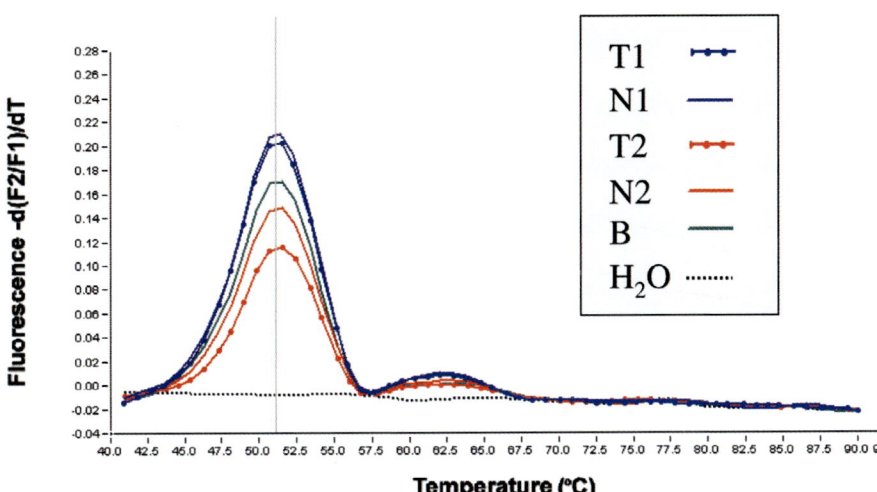

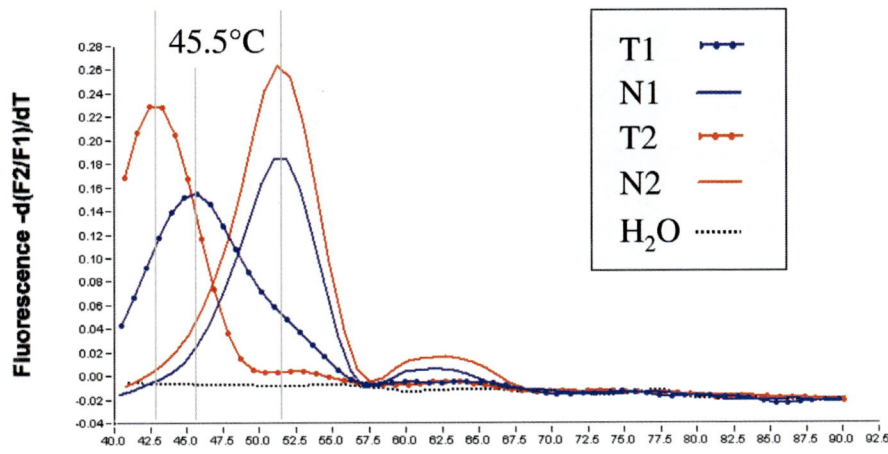

Fig. 1. a BAT26 MSS. Melting peak analysis [-d(F2/F1)/dT] of DNA from two representative MSS tumors (T1, T2), two normal mucosa samples (N1, N2), and a control DNA from mononuclear blood cells (*B*). **b** BAT26 MSI-H. Melting peak analysis [-d(F2/F1)/dT] of DNA from two representative MSI-H tumors (T1, T2) and two normal mucosa samples (N1, N2)

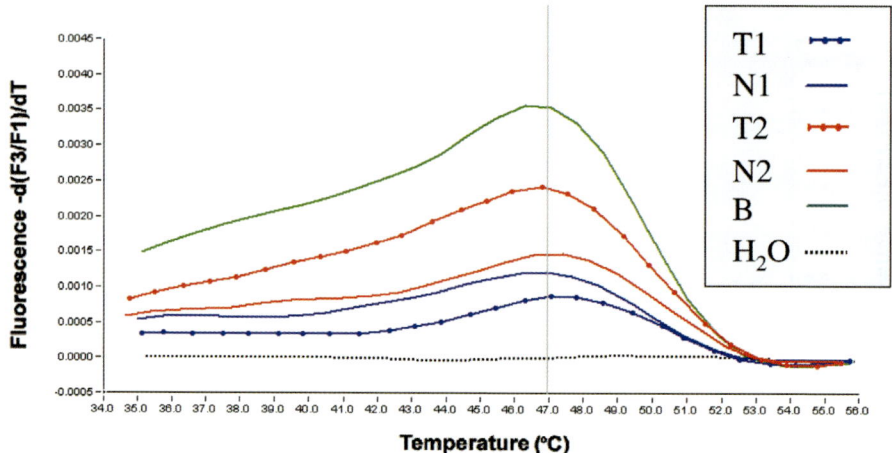

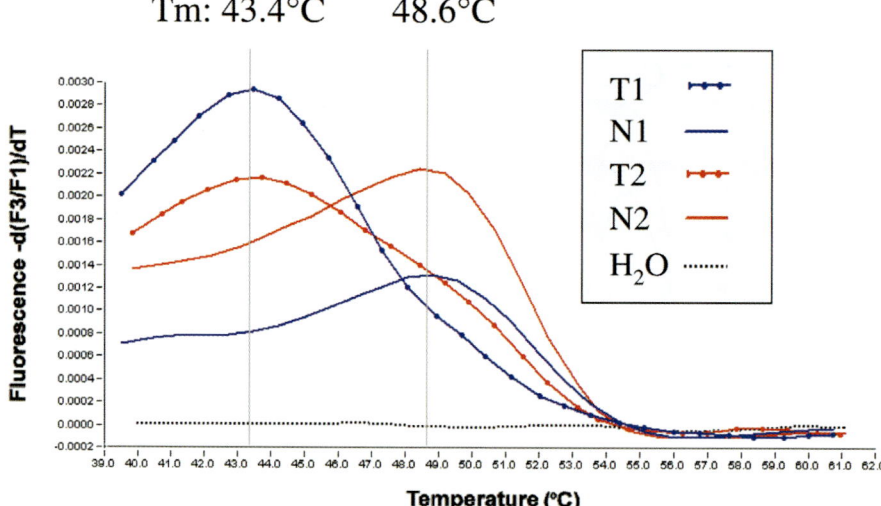

Fig. 2. a BAT25 MSS. Melting peak analysis [-d(F3/F1)/dT] of DNA from two representative MSS tumors (T1, T2), two normal mucosa samples (N1, N2), and a control DNA from mononuclear blood cells (B). b BAT25 MSI-H. Melting peak analysis [-d(F3/F1)/dT] of DNA from two representative MSI-H tumors (T1, T2) and two normal mucosa samples (N1, N2)

(>97%, data not shown) show MSI in both BAT26 and BAT25 markers. Furthermore, it has been reported recently that BAT26 alone can specifically identify all MSI-H tumors from mismatch repair gene mutation-positive CRC patients [14].

The complete microsatellite marker set [5, 6] should be used if no MSI is detectable by BAT26 and BAT25 or only one BAT marker shows MSI, to avoid missing a potential MSI positive tumor.

Due to the differently labeled hybridization probes, it is also possible to perform a duplex BAT26/BAT25 LightCycler PCR, but separate amplification seems to be more robust (data not shown). As seen in our previous work [15], the range of the temperature during the melting point measurement can obviously influence the T_m values. However, there is no significant variation of T_m values of different MSS tumors or normal tissues if the range of melting peak temperature is kept constant for all assays. Since microsatellite melting peak analysis is achieved within an hour of PCR setup, it provides an efficient tool for high-throughput MSI screening analyses.

References

1. Thibodeau SN, Bren G, Schaid D (1993) Microsatellite instability in cancer of the proximal colon. Science 260:81681–81689
2. Fishel R, Lescoe MK, Rao MR, Copeland NG, Jenkins NA, Garber J, Kane M, Kolodner R (1993) The human mutator gene homolog MSH2 and its association with hereditary nonpolyposis colon cancer. Cell 75:1027–3108
3. Aaltonen LA, Peltomaki P, Mecklin JP, Jarvinen H, Jass JR, Green JS, Lynch HT, Watson P, Tallqvist G, Juhola M et al (1994) Replication errors in benign and malignant tumors from hereditary nonpolyposis colorectal cancer patients. Cancer Res 54:1645–1648
4. Lynch HT, Smyrk TC (1998) Identifying hereditary nonpolyposis colorectal cancer. N Engl J Med 338:1537–1538
5. Dietmaier W, Wallinger S, Bocker T, Kullmann F, Fishel R, Ruschoff J (1997) Diagnostic microsatellite instability: definition and correlation with mismatch repair protein expression. Cancer Res 57:4749–4756
6. Boland CR, Thibodeau SN, Hamilton SR, Sidransky D, Eshleman JR, Burt RW, Meltzer SJ, Rodriguez-Bigas MA, Fodde R, Ranzani GN, Srivastava S (1998) A National Cancer Institute Workshop on microsatellite instability for cancer detection and familial predisposition: development of international criteria for the determination of microsatellite instability in colorectal cancer. Cancer Res 58:5248–5257
7. Schlegel J, Vogt T, Munkel K, Ruschoff J (1996) DNA fingerprinting of mammalian cell lines using nonradioactive arbitrarily primed PCR (AP-PCR). Biotechniques 20:178–180
8. Mansfield DC, Brown AF, Green DK, Carothers AD, Morris SW, Evans HJ, Wright AF (1994) Automation of genetic linkage analysis using fluorescent microsatellite markers. Genomics 24:225–233
9. Gyapay G, Ginot F, Nguyen S, Vignal A, Weissenbach J (1996) Genotyping procedures in linkage papping. Methods 9:91–97
10. Papadopoulos N, Nicolaides NC, Wei YF, Ruben SM, Carter KC, Rosen CA, Haseltine WA, Fleischmann RD, Fraser CM, Adams MD, Venter JC, Hamilton SR, Peterson GM, Watson P, Lynch HAT, Peltomäki P, Mecklin JP, de la Chapelle A, Kinzler KW, Vogelstein B (1994) Mutation of a mutL homolog in hereditary colon cancer. Science 263:1625–1629
11. Hoang JM, Cottu PH, Thuille B, Salmon RJ, Thomas G, Hamelin R (1997) BAT-26, an indicator of the replication error phenotype in colorectal cancers and cell lines. Cancer Res 57:300–303

12. Cravo M, Lage P, Albuquerque C, Chaves P, Claro I, Gomes T, Gaspar C, Fidalgo P, Soares J, Nobre-Leitao C (1999) BAT-26 identifies sporadic colorectal cancers with mutator phenotype: a correlative study with clinico-pathological features and mutations in mismatch repair genes. J Pathol 188:252–257
13. Stone JG, Tomlinson IP, Houlston RS (2000) Optimising methods for determining RER status in colorectal cancers. Cancer Lett 149:15–20
14. Loukola A, Eklin K, Laiho P, Salovaara R, Kristo P, Jarvinen H, Mecklin JP, Launonen V, Aaltonen LA (2001) Microsatellite marker analysis in screening for hereditary nonpolyposis colorectal cancer (HNPCC). Cancer Res 61:4545–4549
15. Dietmaier W and Hofstädter F (2001) Detection of microsatellite instability by real time PCR and hybridization probe melting point analysis. LabInvest, 81:1453–1456

Two Color Multiplexing and Typing of Human Papillomavirus Types 16, 18 and 45 on LightCycler

Monica L. Henriquez, Brian E. Caplin*, Randy P. Rasmussen

Introduction

Cervical cancer is the second leading cause of cancer in women worldwide [1,2]. Human Papillomaviruses (HPV) have been identified as the most important viral group associated with benign and malignant neoplasia in humans [3]. Several studies have demonstrated that infection with certain HPV types may progress over a period of years through the various stages of cervical intraepithelial neoplasia (CIN) to invasive squamous carcinoma [4,5,6]. There are more than 80 different HPV types that have been identified [7,8]. Currently, there are more than 20 HPV types that have been found linked to cervical cancer [9]. This group is further subdivided into three categories designated as low, medium and high to differentiate between the relative risk for developing cervical carcinoma from any particular HPV type infection [10]. From the high risk category HPV 16 ranks first followed by HPV 18 in being found at a higher frequency in CIN and cervical carcinomas [1,11]. As of today, HPV 45 is considered a medium/high risk type but there is supporting evidence indicating that HPV 45 follows HPV 18 in frequency and that it should be classified as a high risk type [1, 12].

Unfortunately, to this date, there is no method of detection or characterization that identifies HPV infections by type. The Hybrid Capture I Assay (Digene Corp, Bellsville, MD), the current diagnostic test for the screening of cervical cancer characterizes HPV infections as low risk or medium/high risk but fails to identify HPV infections by type [13].

Polymerase Chain Reaction (PCR) is a highly sensitive method that has been shown to be an effective tool and the most sensitive method for the detection of several HPV types [14, 15,16]. However, analysis of PCR products is for the most part tedious and at many times ambiguous. Conventional methods such as ethidium bromide stained agarose gel electrophoresis may show complicated patterns (multiple bands or smears) making it difficult to interpret gel data [17]. Therefore, the task of designing an assay that will detect and identify HPV infection by type is challenging and difficult.

We have developed a method that uses a homogenous multiplex real-time PCR technique that amplifies and detects HPV 16, HPV 18 and HPV 45 all in one reac-

---

* Brian E. Caplin (✉) (e-mail: brianc@idahotech.com)
  Idaho Technology, 390 Wakara Way, Salt Lake City, UT 84108, USA

tion capillary in approximately an hour. A single primer set amplifies all three templates to assure similar amplification efficiencies. Detection takes place by analysis of product melting profiles [18, 19]. A single fluorescein labeled probe is used to excite two sensor probes, one labeled with LC-Red 640 detects HPV 18 and HPV 45 and a second LC-Red705 labeled probe detects HPV 16. In the case of HPV 18 and HPV 45 that are detected by a single probe, the products can be identified by unique melting profiles. The amplification of the targets cannot be visualized during the run program because the probes have a lower melting temperature than the primers.

Materials

Equipment LightCycler instrument (Roche Diagnostics, Mannheim, Germany)
Software version 3

Reagents Human Papillomavirus Detection kit #4 (IT BioChem, Salt Lake City, USA)
Insta-Mini-Prep Kit (Eppendorf- 5 Prime, Boulder, USA)
High Pure Plasmid Isolation Kit (Roche Diagnostics, Mannheim)
KlenTaq (AB Peptides, St. Louis, USA)
TaqStart Antibody (Clonetech Inc., Palo Alto, USA)
10X Reaction buffer (Idaho Technology Inc., Salt Lake City, USA)
Enzyme Diluent (Idaho Technology Inc., Salt Lake City, USA)
Proteinase K
3M Sodium Acetate pH 5.3
Ethanol (75% and 100%)
10mM Tris pH 8.0
0.1mM EDTA

Procedure

HPV 16 and HPV 18 plasmids were obtained from the American Type Culture Collection (ATCC, Manassas, VA). We could not obtain an HPV 45 plasmid due to its low occurrence in nature. We therefore constructed an artificial HPV 45 template by taking advantage of the sequence similarity between HPV 18 and HPV 45 [12]. The first step consisted of making an HPV45/HPV 18 hybrid sequence by using primers that were specific for HPV 45, the template being HPV 18. The primers used were 45art5' and HPV1300r. The second step attached an EcoRI linker at both ends of the product. The two restriction sites were added for future cloning purposes and were found on the 45art5'2 and 45art3' primers used during this step. These steps altogether produced a template with primer and probe annealing sites specific for HPV 45 (see figure 1). Six additional HPV plasmids were obtained from ATTC. One plasmid of no risk (HPV2), three plasmids of low risk (HPV 6B, HPV 11 and HPV 44) and two plasmids of medium risk (HPV 31 and HPV 56) were used as controls in this study [10]. Plasmids were extracted

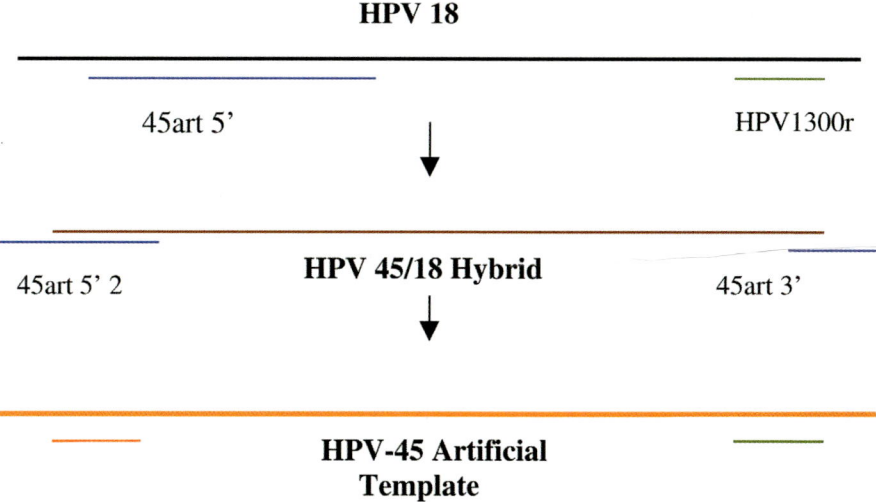

Fig. 1. Construction of an artificial HPV-45 required two amplification steps. The first step produced an HPV45/HPV-18 hybrid sequence. The second consisted in the attachment of EcoRI linkers to produce the full artificial HPV-45 template. These steps produced a template that has primer annealing sites for HPV-45

with the Insta-Mini-Prep Kit from Eppendorf- 5 Prime (Boulder, CO) and the High Pure Plasmid Isolation Kit (Mannheim, Germany). Extraction procedures were described in the kits. For purification purposes plasmids were Ethanol precipitated.

Thirty-nine clinical samples that had tested positive for the Digene Hybrid Capture I test (Digene Corp, Bellsville, MD) were DNA extracted using standard

Table 1. Oligonucleotides

(GenBank Accession # 74479)				
Primers	Position	Length	GC (%)	T_m (°C)
45art 5' GATCCAGAAGGTACCGACGGGGAGGGAACG- GGGTGTAATGGCTGGTTCTTTGTAGAAACA- ATTGTAGAGAAAAA		74	47.3	82.9
HPV1300r CCACTTCAGAATTGCCATAGCC		22	50.0	63.6
45art 5' 2 GGGAATTCCAGTAAGCAACAATGGCGGATC- CAGAAGGTACCGACGGG		47	55.3	81.75
45art 3' GGCTTAAGCTTCCACTTCAGAACAGCCATAGCCA		34	50.0	75.7

phenol-chloroform procedure. For purification purposes, samples were Ethanol precipitated using traditional procedure. Following Ethanol precipitation, each pellet was resuspended in 100μl of 1X TE' (10mM Tris pH 8.0, 0.1mM EDTA), vortexed vigorously and transferred to a 1.5ml microfuge tube. The DNA was boiled in a water bath at 98°C for 10 minutes with the microfuge lid opened. Samples were cooled and placed in the freezer at –20°C until tested.

The high risk amplification primers and probes were obtained from the Human Papillomavirus Detection kit #4.

The following master mix was used for amplification and detection of high risk HPV types in a total volume of 20μl per reaction:

	Volume [μl]	[Final]
10X 35.5mM MgCl$_2$ reaction buffer	2	3.5mM
dUTPs	2	2
Taq Polymerase*	2	0.8U
TaqStart Antibody*		0.2μg
HPVrev**	2	0.5uM
HPV FITC**	2	0.5uM
HPV 18/45 LCRed 640**	2	0.2uM
HPV 16 LCRed 705**	2	0.2uM
Human Genomic DNA	1	50ng/ul
HPV sample	1	
water	4	

* The KlenTaq/TaqStart mixture was made by combining at room temperature 2μl of KlenTaq (AB Peptides, USA) with 2μl of TaqStart Antibody (Clonetech, USA) with an incubation time of 5 minutes. The KlenTaq/TaqStart mixture was diluted into 21 μl of Taq Diluent (Idaho Technology, USA).
** Human Papillomavirus Detection kit #4 (IT Biochem, Salt Lake City, USA)

18μl of master mix plus DNA were transferred to a capillary tube, sealed and centrifuged. Capillaries were placed in the LightCycler carousel.

- Initial denaturation and activation of enzyme in FastStart kit occurred for 10 min at 95°C
- Amplification

Parameter	Value		
Cycles	40		
Type	Quantification		
	Segment 1	Segment 2	Segment 3
Target temperature [°C]	95	62	72
Incubation time [s]	0	10	9
Temperature transition rate [°C/s]	20	10	20
Acquisition mode	None	Single	None
Gains	F1=1	F2=10	F3=19

- Melting Curve Analysis

Parameter	Value		
Cycles	1		
Type	Melting curves		
	Segment 1	Segment 2	Segment 3
Target temperature [°C]	95	48	70
Incubation time [s]	10	180	0
Temperature transition rate [20°C/s]	20	5	0.2
Acquisition mode	None	None	Step

- Color compensation was used to correct signal bleed through of channels 1 and 2 into channel 3.

Analysis

The melting curve profile for each product was analyzed using the "Melting Curve Analysis" format of the LightCycler Data Analysis (LCDA) software. The "Melting Curve Analysis" format uses unique algorithms that take the negative derivative of the product melting curve generating in turn melting peaks. Once in this program, there are certain parameters that must be set for the detection of HPV high risk types.

- The Melting Curve calculation needs to be set as "polynomial with background subtraction" in both channels 2 and 3 to compensate for the non-specific effects of temperature on fluorescence.
- The background subtraction cursors (2 green and 2 blue vertical lines) in F2 need to be placed at the following temperatures (°C): 51.00, 52.00, 64.00 and 66.00. The cursors in F3 need to be set at the following temperatures (°C): 53.00, 54.00, 66.00 and 68.00.
- "°C to Average" which sets the temperature interval used to calculate the derivative data should be placed at a value of 8.0.

Results

Detection of HPV 16, HPV 18 and HPV 45

Two color multiplexing and analysis of HPV 16, HPV 18 and HPV 45 was accomplished on the LightCycler instrument. HPV 18 and HPV 45 were detected by the HPV 18/45 LC-Red 640 sensor probe and identified in channel 2 based on melting temperature. The HPV 18 product had a melting temperature of about 59.57°C while the HPV 45 product had a melting temperature of 56.06°C (see figures 2). The 4 C difference in melting temperature was used to differentiate these two closely related HPV high risk types making the analysis fast, uncomplicated and reliable. The HPV 16 product was detected by the LC-Red 705 sensor probe and seen in channel 3. The melting temperature of the HPV 16 product appeared to be approximately 60.66°C (see figure 2). In summary, the HPV 18 and HPV 45 products can be differentiated by melting temperature while HPV 18 and HPV 16 products can be distinguished by color. The HPV 16 and HPV 45 products can be distinguished both by melting temperature and color.

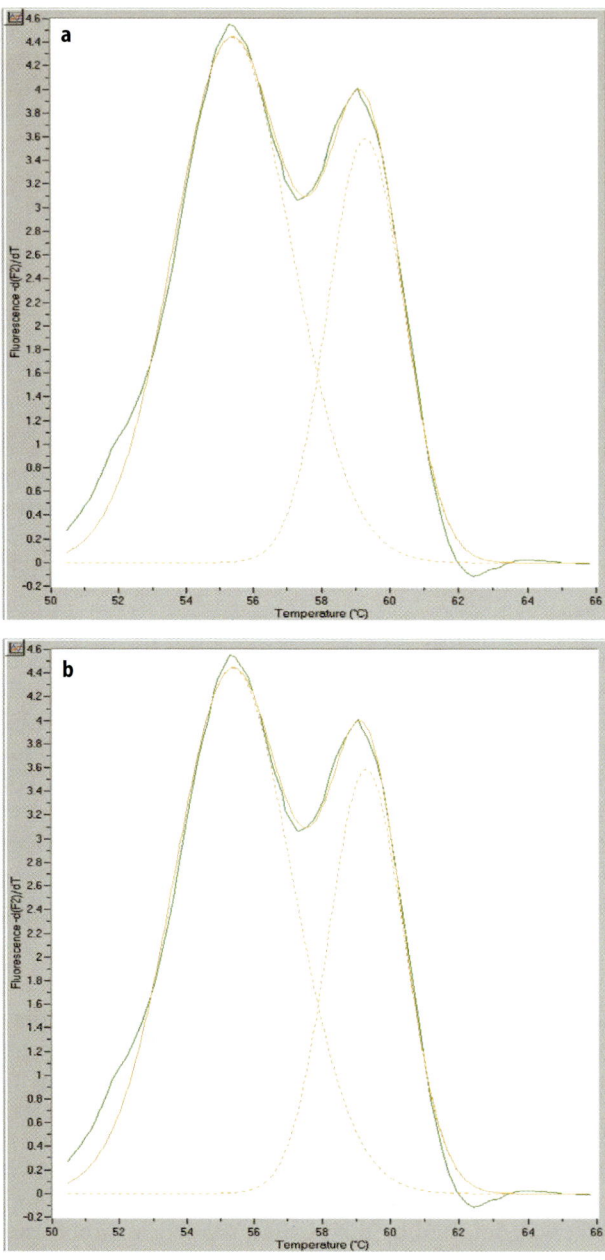

Fig. 2. Two color multiplexing of HPV 16, HPV 18 and HPV 45 in a single capillary. The starting concentration of each subtype was approximately 10^5 copies/μl. All three subtypes were amplified in a background of 50ng/μl of human genomic DNA. (**a**) The HPV 18 and HPV 45 products were detected in channel 2 and differentiated by melting temperature. (**b**) HPV 16 was characterized by color in channel 3

Specificity of the Detection System

To test the specificity of the probes, nine HPV types served as controls and were amplified separately with the same primer set in the presence of the anchor and the two sensor probes. Each HPV control had a starting concentration of 10^5 copies/ìl and was amplified in a background of 50ng/ìl of human genomic DNA. The results demonstrated that the HPV 16, HPV 18 and HPV 45 targets were amplified and detected. The three low risk HPV types 6b, 11 and 44 as well as HPV 2 (no risk type) were not detected. HPV 31 and HPV 56 that are considered medium risk types generated specific melting curve profiles that had the same melting temperature in both channels 2 and 3 (see figure 3).

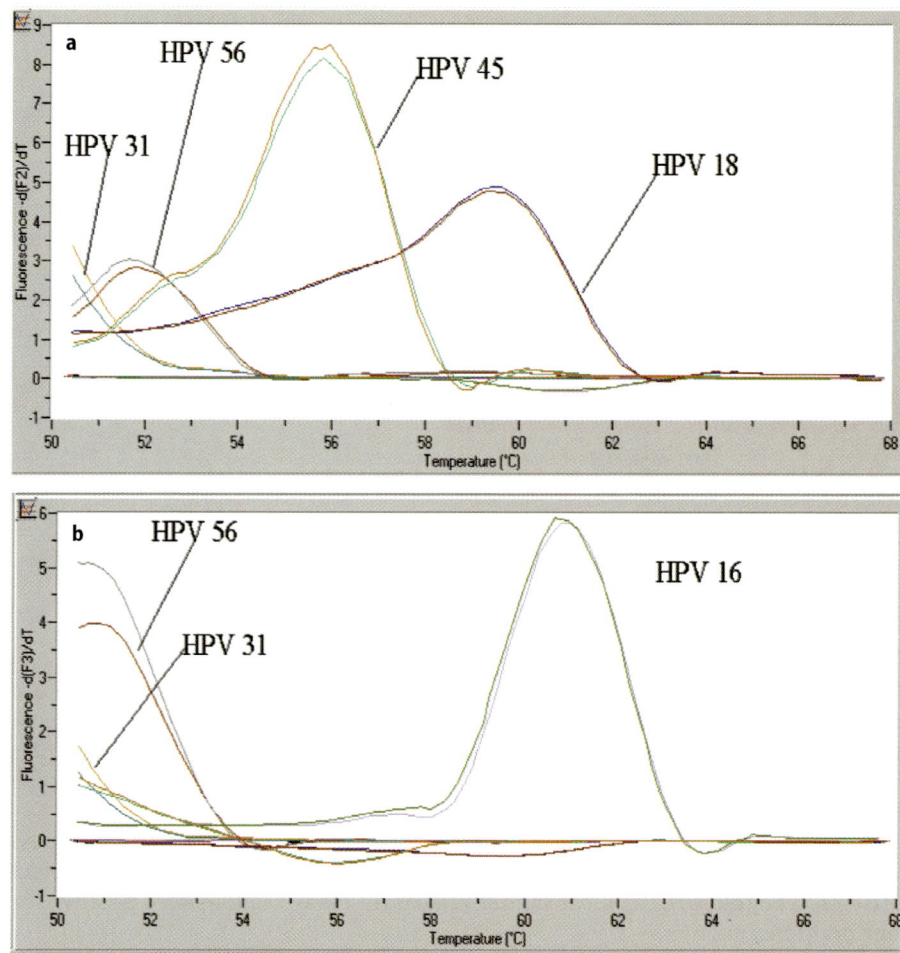

Fig. 3. Data was not background subtracted. (**a**) Medium risk HPV types 31 and 56 detected in channel 2. (**b**) Medium risk types 31 and 56 detected in channel 3. HPV 31 had the same melting temperature in channels 2 and 3 (approximately 48°C) and 56 had nearly the same melting temperature (about 51.5°C) channels 2 and 3

Table 2. Summary of the nine HPV controls types tested for the specificity of the detection system

HPV type	Category	Detected	Comment
2	No risk	No	–
6b	Low risk	No	–
11	Low risk	No	–
16	High risk	Yes	Product T_m 60.66°C (F3)
18	High risk	Yes	Product T_m 59.57°C (F2)
31	Medium risk	Yes	T_m 48°C (F2, F3)
44	Low risk	No	–
45	High risk	Yes	Product T_m 56.06°C (F2)
56	Medium risk	Yes	T_m 51.82°C (F2, F3)

Limit of Detection for HPV 16, HPV 18 and HPV 45

The limit of detection for all the high risk HPV types can be found on the table below. HPV types were amplified in a background of 50ng/µl of human genomic DNA. The limit of detection for each type was determined based on the presence of melting peaks since the amplification cannot be visualized in real time because the probes have a lower melting temperature than the primers.

N=8

HPV type	Limit of Detection
16	1000 copies/µl
18	100 copies/µl
45	100 copies/µl

Testing of Clinical Samples

A total of 39 clinical samples that had been screened with the Digene Capture Hybrid I assay were tested with the multiplex assay to see how well the results correlated with each other. The Digene Capture Hybrid I assay characterizes samples into two categories: low risk type (A probe positive) and medium/high risk type (B probe positive) but it does not classify samples by type.

From the 39 samples, the multiplex assay identified six samples as HPV 16 positive and one was characterized as HPV 18 positive. We were not able to detect HPV 45. The following clinical samples listed on table 3 below were positive for our multiplex assay.

The multiplex assay results seem to correlate well with the Digene Hybrid Capture I assay. All the B+ probe samples were either high risk or medium risk HPV types by the multiplex assay. Sequencing data also correlated with the multiplex assay results.

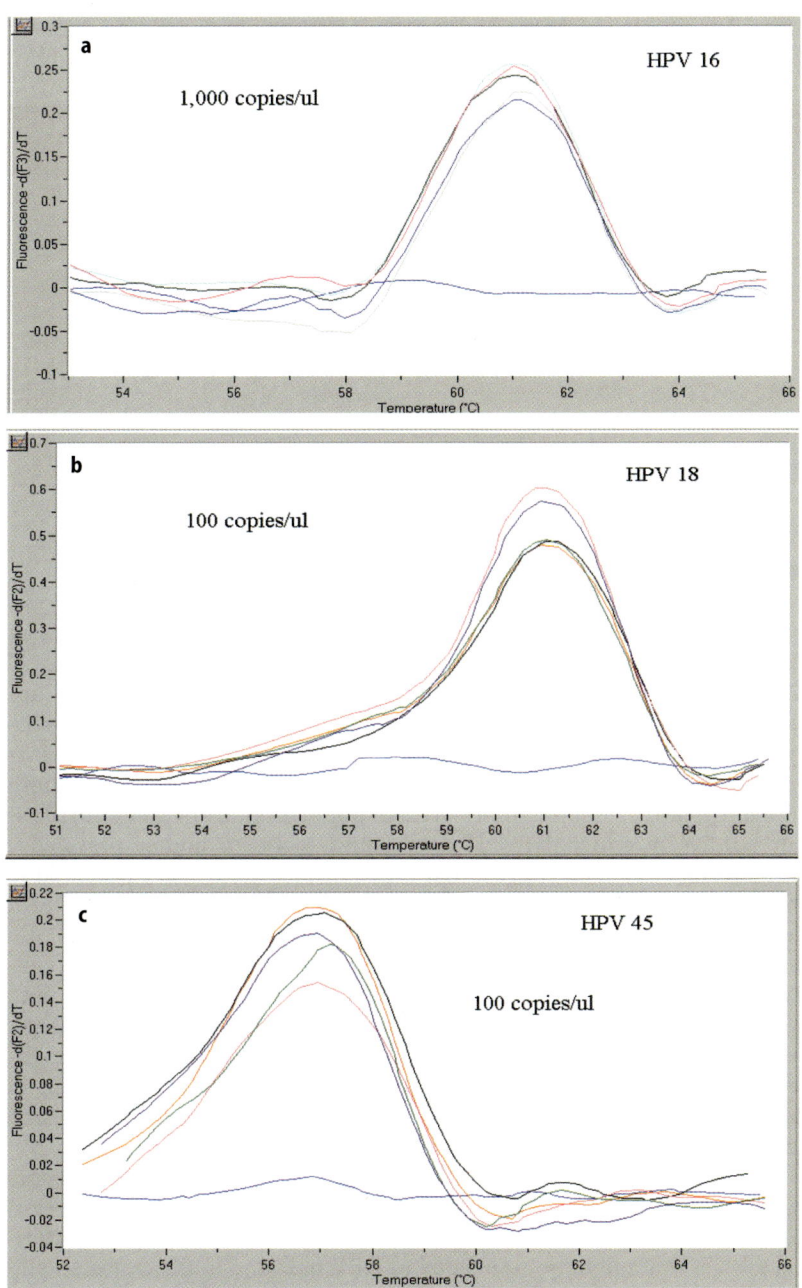

Fig. 4: Graphs showing the limit of detection for each high risk HPV type. Amplification of each HPV type took place in the presence of 50ng/μl of human genomic DNA. (**a**) The limit of detection for HPV 16 was 1,000 copies/μl. (**b**) The limit of detection for HPV 18 was 100 copies/μl. (**c**) The limit of detection for HPV 45 was 100 copies/μl

Table 3. Results of 39 clinical samples tested. The melting temperature (T_m) of HPV high risk controls were: HPV 16 = 60.66°C; HPV 18 = 59.57°C; HPV 45 = 56.06°C. This table compares Digene Hybrid Capture I assay with multiplex assay results

Name	Tm F2	Tm F3	HPV type	Digene Capture Hybrid I assay	Sequencing data
1. Sample 1		60.69°C	HPV 16	B+	–
2. Sample 2*	~47°C	~47°C	Medium risk	A+B+	–
3. Sample 4		61.15°C	HPV 16	B+	HPV 16
4. Sample 5		60.75°C	HPV 16	A+B+	–
5. Sample 6	59.85°C		HPV 18	B+	HPV 18
6. Sample 11		60.67°C	HPV 16	A+B+	
7. Sample 27		60.88°C	HPV 16	B+	–
8. Sample 38		60.58°C	HPV 16	B+	HPV 16

*Data was not background subtracted

Discussion

In our research, we have been able to demonstrate a multiplex PCR assay in which HPV 16, HPV 18 and HPV 45 have been specifically characterized. In our small study of 39 clinical samples that had been tested with the Digene Hybrid Capture I assay, 15.4% were found to be HPV 16 positive and 2.5 % were HPV 18 positive. We were not able to detect HPV 45 infected samples perhaps because it is mostly found in HIV positive women or immunocompromised patients and both groups were not well represented in our sample population. Our results corroborated other studies in which HPV 16 was found to be in higher prevalence [1,10,16]. The results obtained suggest that the multiplex assay has a potential to be used in a clinical setting in addition to the Digene Hybrid Capture I test for it can provide vital information to both the clinician and the patient.

Due to the degeneracy of the primer and probe sets some of the medium types were also detected by our assay. These types were simultaneously detected in both channels. This is probably the result of the anchor probe melting off the target before the sensor probe which explains why the medium risk have a lower melting temperature. However, if the correct analysis parameters are set according to the instructions previously explained, distinguishing between high risk and medium risk HPV types should be uncomplicated.

This assay has not yet been validated for the detection of high risk HPV types since further testing is required. All the HPV types that are linked to cervical cancer should be tested to check for assay specificity. In addition, a panel of bacterial and viral species that might be associated to HPV or found in the cervix should be tested to check for assay possible cross-reactivity.

References

1. Bosch FX, Mano MM, Munoz N, Sherman M, Jansen AM, Peto J, Schiffman MH, Moreno V, Kurman R, Shah KV (1995) Prevalence of human papillomavirus in cervical cancer: a worldwide perspective. J National Cancer Inst 87(11):796–802
2. Walboomers JMM, Jacobs MV, Manos MM, Bosch FX, Kummer JA, Shah KV, Snijders PJF, Peto J, Meijer CJLM, Munoz N (1999) Human papillomavirus is a necessary cause of invasive cervical cancer worldwide. J Pathology 189:12–19
3. Chan SY, Delius H, Halpern AL, Bernard HU (1995) Analysis of genomic sequences of 95 papillomavirus types: uniting typing, phylogeny, and taxonomy. J of Virology 69(5):3074–3083
4. White DO, Fenner FJ (1994) Human Papillomavirus. Med Virology 4th Edition: 297–298
5. Schiffman MH, Bauer Hm, Hoover RN, Glass AG, Cadell DM, Rush BB, Scott DR, Sherman ME, Kurman RJ, Wacholder S, Stanton CK, Manos MM (1993) Epidemiologic evidence showing that human papillomavirus infection causes most cervical intraepithelial neoplasia. J of Natl Cancer Inst 85(12):958–964
6. Ho GYF, Burk RD, Klein S, Kaddish AS, Chang CJ, Palan P, Basu J, Tachezy R, Lewis R, Romney S (1995) Persistent genital human papillomavirus infection as a risk factor for persistent cervical dysplasia. J of Natl Cancer Inst 87(18):1365–1371
7. Hart KW, Williams OM, Thelwell N, Fiander AN, Brown t, Borysiewicz LK, Gelder CM (2001) Novel method for detection, typing, and quantification of human papillomaviruses in clinical samples. J of Clin Microbiology 39(9):3204- 3212.
8. Menzo S, Monachetti A, Trozzi C, Ciavattini A, Carloni G, Varaldo P, Clementi M (2001) Identification of six putative novel human papillomaviruses (HPV) and characterization of candidate HPV type 87. J of Virology 75(23):11913–11919.
9. Zheng PS, Li SR, Iwasaka T, Song J, Cui MH, Sugimori H (1995) Simultaneous detection by consensus multiplex PCR of high- and low- risk and other types of human papilloma virus in clinical samples. Gynecologic Oncology 58:179–183
10. Shah KV and Howley PM (1996) Papillomaviruses. Fields Virology 3rd Edition :2077–2109
11. Roden RBS, Greenstone HL, Kirnbauer R, Booy FP, Jessie J, Lowy DR, Schiller JT (1996) In vitro Generation and type-specific human papillomavirus type 16 virion pseudotype. J of Vir 70(9):5875–5883
12. van Muyden RCPA, Harmsel BWA, Smedts FMM, Hermans J, Kuijpers JC, Raikhlin NT, Petrov S, Lebedev A, Ramaekers FCS, Trimbos JB, Kleter B, Quint WGV (1999) Detection and typing of human papillomavirus in cervical carcinomas in Russian women. Cancer 85(9):2011–2016
13. Clavel C, Masure M, Bory JP, Putaud I, Mangeonjean C, Lorenzato M, Gabriel R, Quereux C, Birembaut (1999) Hybrid capture II-based human papillomviruse detection, a sensitive test to detect in routine high-grade cervical lesions: a preliminary study on 1518 women. Brit J Cancer 80(9):1306–1311
14. Poljak M, Seme K (1996) Rapid detection and typing of human papillomavirus by consensus polymerase chain reaction and enzyme-linked immunosorbent assay. J Vir Methods 56:231–238
15. Tieben LM, Schegget Jt, Minnaar RP, Bavinck JNB, Berkhout RJM, Vermeer BJ, Jebbink MF, Smits HL (1993) Detection of cutaneous and genital HPV types in clinical samples by PCR using consensus primers. J Vir Methods 42:265–280
16. Ferenczy A (1995) Viral testing for genital human papillomavirus infections: recent progress and clinical potentials. Int J Gynecol Cancer 5:321–328
17. Godfroid E, Heinderyckx M, Mansy F, Fayt I, Noel JC, Thiry L, Bollen A (1998) Detection and identification of human papilloma viral DNA, types 16, 18, and 33, by a combination of polymerase chain reaction and a colorimetric solid phase capture hybridization assay. J of Vir Methods 75:69–81
18. Ririe KM, Rasmussen RP, Wittwer CT (1997) Product differentiation by analysis of DNA melting curves during the polymerase chain reaction. Analytical Biochem 244:154–160
19. Lay MJ, Wittwer CT (1997) Real-time fluorescence genotyping of factor V leiden during rapid-cycle PCR. Clin Chem 43(12):1–6

Quantitative Analysis of AML1-ETO Fusion Transcripts in t(8;21) Positive AML Using Real-Time RT-PCR

MARTIN WEISSER*, CLAUDIA SCHOCH, TORSTEN HAFERLACH, WOLFGANG HIDDEMANN, SUSANNE SCHNITTGER

Introduction

Acute myeloid leukemia (AML) is an aggressive hematologic malignancy that requires urgent cytostatic treatment. The incidence of AML is approximately 2–4 new cases per 100,000 with a median age of about 65 years. Despite intensive treatment with cytotoxic drugs and bone marrow transplantation, long-term survival is less than 30% [1, 2]. In AML an early progenitor is transformed to a leukemic blast that proliferates in blood and bone marrow and suppresses normal hematopoiesis. In about 55% of cases the leukemic blasts show cytogenetic abnormalities that may be associated with distinct biological and clinical features. These specific cytogenetic abnormalities can be detected via laboratory techniques like classical cytogenetics, FISH, Southern blotting or PCR. The specific karyotype of the leukemic clone is not only one of the most important prognostic factors in AML, but it also facilitates monitoring of minimal residual disease at times of cytomorphologic remission when leukemic blasts may be undetectable by light microscopy. At primary diagnosis the estimated leukemic burden is about 10^{12} malignant cells. Complete remission after chemotherapy is defined as the return to normal bone marrow cytomorphology with less than 5% myeloid blasts. However, in the state of cytomorphologic remission patients may still have 10^{10} leukemic blasts and relapse is a common cause of treatment failure [4]. The kinetics of leukemic regrowth and host factors like the immune response that influence residual leukemia remain largely unknown.

One of the most frequent chromosomal aberrations in AML is the translocation t(8;21). In t(8;21) positive AML, the AML1 gene on chromosome 21 is fused to the ETO gene on chromosome 8. The AML1 gene encodes a DNA binding factor while the function of the ETO gene remains unknown [5]. This leads to the expression of a constant fusion transcript that can be detected by RT-PCR [6] (Fig. 1). In all published studies t(8;21) is associated with a relatively good prognosis compared to other AMLs with normal cytogenetics or complex karyotypes. However, 10–20% of the patients still relapse after conventional chemotherapy [7]. Using novel quantitative real-time RT-PCR techniques we

---

* Martin Weisser (✉) (e-mail: martin.weisser@med3.med.uni-muenchen.de)
  Medizinische Klinik III, Klinikum Großhadern,
  Marchioninistr. 15, 81377 München

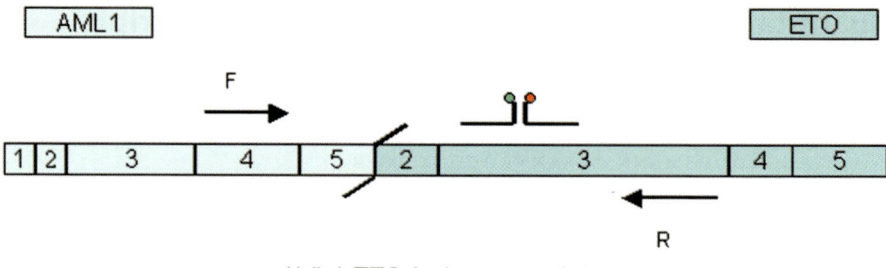

Fig. 1. Schematic diagram of the AML1-ETO transcript with AML1 exon 5 fused to ETO exon 2. The breakpoint region and the position of the forward (*F*) and reverse (*R*) primers and the hybridization probes are indicated

established a sensitive method to accurately quantify minimal residual leukemia. Quantification of AML1-ETO transcript levels may lead to the detection of a critical MRD level or via disease kinetics (increase and decrease of transcript levels) reveal patients with a high risk of hematological relapse before clinical manifestation.

Materials

Equipment LightCycler instrument (Roche Diagnostics, Mannheim, Germany)
LightCycler software 3.0 and a preliminary-version of the relative quantification software (kindly provided by Dr. R. Soong, Roche Diagnostics, Penzberg)
LightCycler capillaries and centrifuge (Roche Diagnostics)
EDTA stabilized human blood and bone marrow samples
Kasumi 1 cell line [8].

Reagents LightCycler Faststart DNA Master Hybridization Probes (Roche Diagnostics)
1st Strand cDNA Synthesis Kit for RT-PCR (AMV) (Boehringer, Mannheim, Germany)
Qiagen RNeasy RNA-Isolation Kit (Qiagen, Hilden, Germany)
Hybridization Probes (TIB MOLBIOL, Berlin, Germany)
Primers (TIB MOLBIOL)
Ficoll Hypaque density gradient (Sigma, Munich, Germany)

Procedure

Human blood and bone marrow samples were analyzed. Patients were selected due to t(8;21) positive cytogenetics and were referred to our lab for central AML diagnosis. Mononuclear cells were obtained by Ficoll Hypaque density gradient centrifugation. Total RNA of 1×10^7 human cells and Kasumi 1 cells was isolated according to the Qiagen RNeasy protocol. cDNA Synthesis was performed via Boehringer Mannheim 1st Strand cDNA Synthesis Kit for RT-PCR (AMV) using random hexamers, according to the manufacturer's protocol. cDNA of 10^6 cells was then amplified using the LightCycler Faststart DNA Master. The PCR products were detected via specific fluorescence labeled hybridization probes. The detection format is based on Fluorescence Resonance Energy Transfer (FRET) [9, 10]. To measure the sensitivity of our method we performed limiting dilution experiments of Kasumi 1 cells in the background of mononuclear cells of a healthy donor.

Analysis of LightCycler Data

LightCycler data was analyzed using LightCycler 3.0 software and the second derivative maximum method. Quantitative analysis was performed relative to the housekeeping genes G6PDH and cABL [11, 12]. Therefore the ratio of target gene to housekeeping gene was calculated. Relative quantification of AML1-ETO fusion transcripts was calculated according to standard curve analysis and in parallel using a known concentration of an external calibrator. We used cDNA of 10^5 Kasumi 1 cells as an external calibrator and analyzed the transcript levels of the unknown samples by using a β-version of the relative quantification software. The AML1-ETO level of the calibrator sample (10^5 Kasumi 1 cells) was set to a relative value of 1,000,000. PCR efficiencies for target and housekeeping genes were measured by repeated generation of standard curves using dilution series of Kasumi 1 cells and these efficiencies were used to calculate unknown sample concentrations.

PCR was performed using 2 µl mastermix (LightCycler Faststart DNA Master Hybridization Probes, Roche Diagnostics, Mannheim, Germany), containing buffer, dNTPs and Taq polymerase; 4 mM MgCL2; 0.25 µM of each 3′ and 5′ fluorescent labeled hybridization probes (TIB Molbiol, Berlin, Germany); 0.5 µM of each 3′ and 5′ Primer (TIB Molbiol, Berlin, Germany), 2 µl of cDNA and H_2O to a final volume of 20 µl. The reaction mix for AML1-ETO, G6PDH and cABL amplification was:

	Volume [µl]	[Final]
LightCycler Faststart DNA Master	2	1X
MgCl$_2$ (25 mM)	2.4	4 mM
Probes (2.5 µM each)	2+2	0.25 µM each
Primers (10 µM each)	1+1	0.5 µM each
cDNA	2	
H$_2$O	7.6	
Total	20	

The PCR protocol was:

Parameter	Value		
Cycles	45		
Type	Quantification		
	Segment 1	Segment 2	Segment 3
Target temperature [°C]	95	64	72
Incubation time [s]	10	10	26
Temperature transition rate [°C/s]	20	20	2
Acquisition mode	None	None	Single
Gains	F1=1; F2=15		

Table 1. Oligonucleotides

AML1-ETO (GenBank Accession #D13979)					
	Position	Exon	Length	GC (%)	T_m(°C)
Primers					
F5'GAGGGAAAAGCTTCACTCTG3'	2,003	4	20	50	60.7
R5"TCGGGTGAAATGTCATTGCC3'	2,450	3	20	50	63.5
Probes					
5'CCCTCACCACCCAATGGCTTCAGC-F	2,293	3	24	62.5	73.3
5'LCRED640-TGGGCCTTCCTCTTCTTCC TCCTCC-P	2,319	3	25	60	72.6
G6PDH (GenBank Accession #X55448)					
Primers					
5'CCGGATCGACCACTACCTGGGCAAG-3'	15,116	G6PDH 6	25	64	73.8
5'GTTCCCCACGTACTGGCCCAGGACCA-3'	16,443R	G6PDH 9	26	65	76.8
Probes					
5'GTTCCAGATGGGGCCGAAGATCCT GTTG-F	15,376R	G6PDH 7	28	57	72.7
5'-LCRED640-CAAATCTCAGCACCAT GAGGTTCTGCAC-P	15,165R	G6PDH 7/6	28	50	70.0
cABL (GenBank Accession #U07563)					
Primers					
5'CCCAACCTTTTCGTTGCACTGT3'	49,996	ABLa2	22	50	66.5
5'CGGCTCTCGGAGGAGACGTAGA3'	58,610R	ABLa4	22	64	69.3
Probes					
5'TGAAAAGCTCCGGGTCTTAGGCTA TAATCA-F	50,627	ABLa3	30	43	70.1
5'-LCRED640-AATGGGAATGGTGTG AAGCCCAAA-P	50,658	ABLa3	25	48	70.7

Results

To access the sensitivity of our real-time PCR protocol we performed a limiting dilution series of a single Kasumi 1 cell in the background of 10, 100, 1000, 10,000, 100,000 1,000,000 normal human cells. This revealed a sensitivity of 10^{-5} (1 Kasumi 1 cell in the background of 100,000 normal cells was detectable) (Fig. 2).

Sensitivity of Real-Time RT-PCR Protocol

Repeated standard curve generation using serial dilutions of Kasumi 1 cells (10^5, 10^4, 10^3, 10^2) was carried out. PCR efficiencies for the amplification of AML1-ETO and G6PDH were calculated using the following formula:

PCR Efficiency

$$\text{PCR efficiency} = 10^{-1/\text{slope}}$$

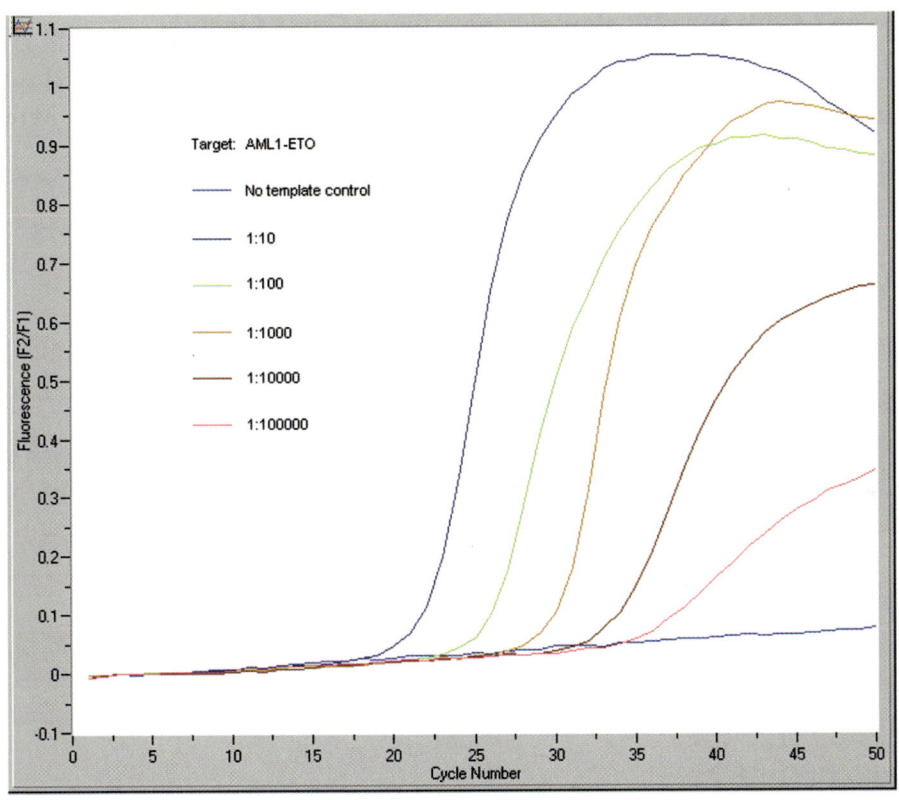

Fig. 2. AML1-ETO transcripts of 1 t(8;21) positive Kasumi 1 cell in the background of 10, 100, 1000, 10,000 and 100,000 normal human cells were detected

The PCR efficiencies for AML1-ETO and G6PDH were 2.03 and 2.01 respectively (Fig. 3).

Comparison of Standard Curve Analysis Versus External Calibrator (Preliminary Version of the Relative Quantification Software)

Comparison To compare the results of standard curve analysis and the external calibrator method, exemplary samples were analyzed in parallel. Quantitative levels of AML1-ETO obtained with either method correlated well (Fig. 4).

Quantification of AML1-ETO Fusion Transcript Levels of a Patient During Early Phase of Induction Chemotherapy Using G6PDH and cABL Housekeeping Genes Simultaneously

Quantification Figure 5 demonstrates the decline of AML1-ETO levels of a patient at primary diagnosis and during induction chemotherapy. Peripheral blood samples were

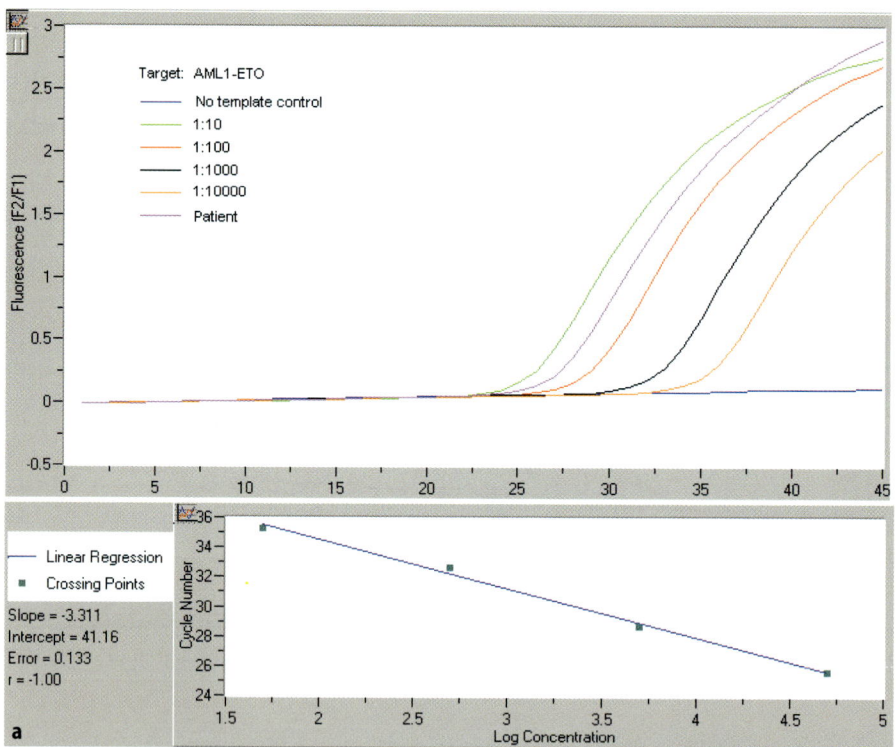

Fig. 3a. Standard curve of AML1-ETO (**a**) and G6PDH (**b**) using serial dilutions of Kasumi 1 cells (1:10 – 1:10000) plus quantitative analysis of a patient

taken twice a week. AML1-ETO was normalized to the housekeeping genes G6PDH and cABL. The relative quantification of AML1-ETO/G6PDH and AML1-ETO/cABL at each point of analysis during induction therapy revealed virtually identical results. Thus G6PDH as well as cABL are appropriate reference genes for AML1-ETO quantification. This confirms data described previously for quantitative analysis of BCR-ABL fusion transcripts in CML [10].

Quantification of AML1-ETO levels using real-time RT-PCR during the course of therapy of a patient demonstrates that real-time PCR has the potential to diagnose molecular relapse before it is detectable by cytogenetics or cytomorphology (Fig. 6).

Quantitative Real-Time PCR has the Potential to Predict Hematological Relapse Before Clinical Manifestation

Conclusions

Quantitative monitoring of AML1-ETO fusion transcript levels in t(8;21) positive AML using real-time PCR offers several advantages. Compared to competitive

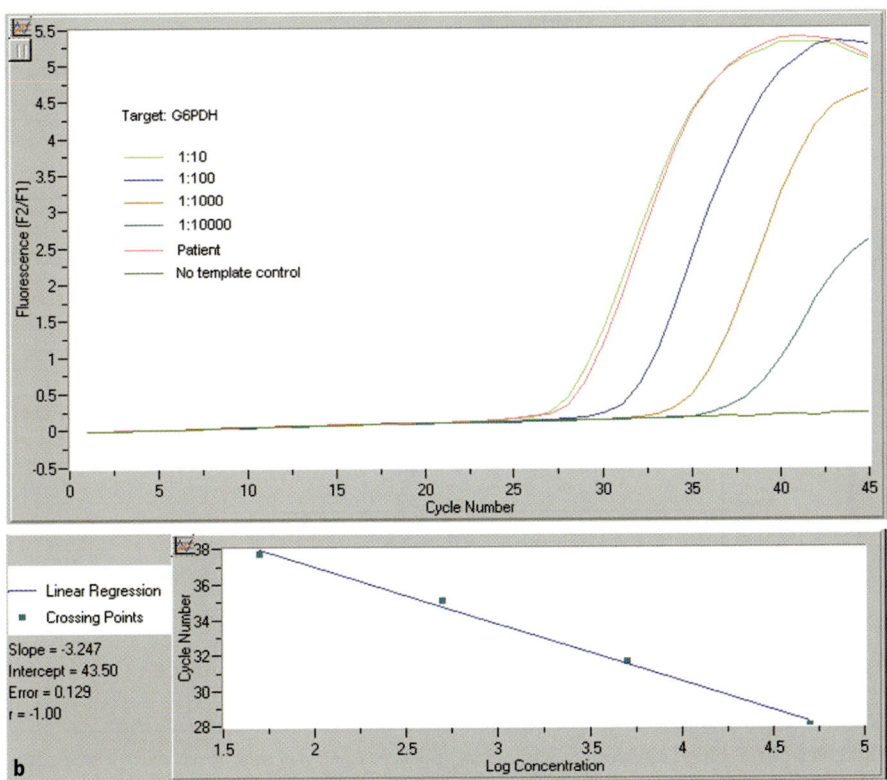

Fig. 3b. Standard curve of AML1-ETO (**a**) and G6PDH (**b**) using serial dilutions of Kasumi 1 cells (1:10 – 1:10000) plus quantitative analysis of a patient

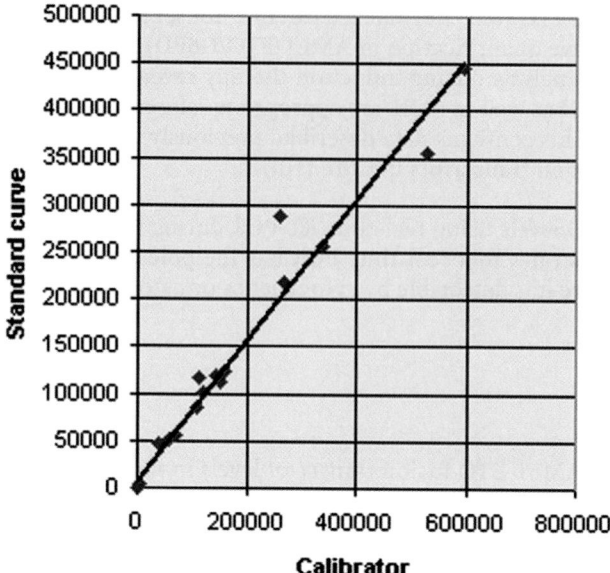

Fig. 4. Correlation of AML1-ETO levels quantified by standard analysis (x-axis) and preliminary version of the relative quantification software (y-axis). Coefficient of correlation =0.98

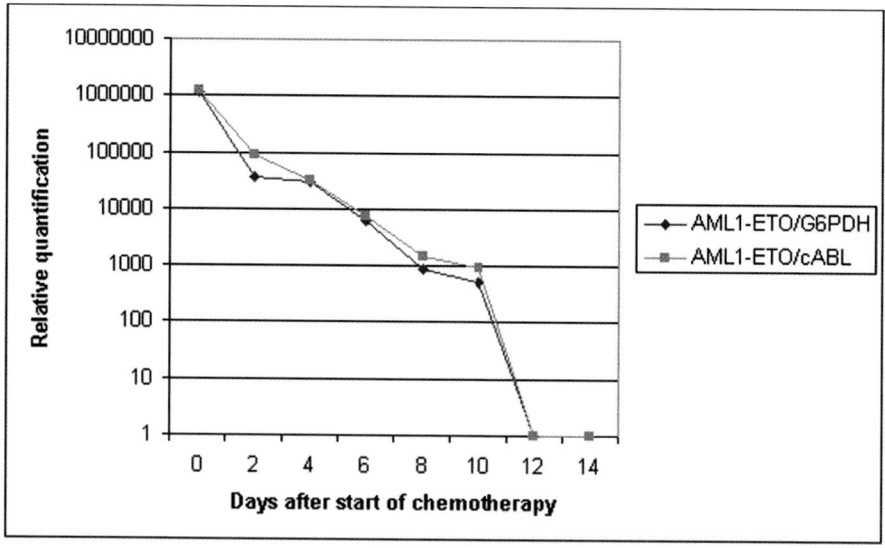

Fig. 5. Peripheral blood of a patient with t(8;21) positive AML was monitored twice a week during induction chemotherapy. AML1-ETO levels using G6PDH and cABL as control genes correlated well. On day 16 after the start of therapy, AML1-ETO was undetectable in peripheral blood samples

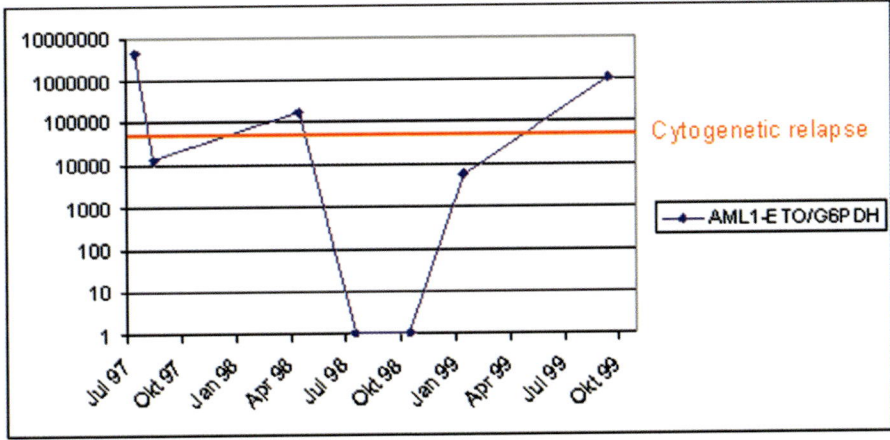

Fig. 6. AML1-ETO levels of a patient were analyzed at different stages of therapy. After induction therapy this patient remained AML1-ETO positive; 5 weeks later he had a cytogenetic relapse. After re-induction therapy AML1-ETO transcripts were undetectable using real-time RT-PCR. He remained AML1-ETO negative for about 2 months. Increasing levels of AML1-ETO were detectable via real-time RT-PCR 5 months before diagnosis of a second cytogenetic relapse

PCR, which is rather time consuming, real-time PCR can be performed in about 1 h. The risk of contamination is reduced as real-time PCR requires no post-PCR sample preparation. As detection of PCR products occurs during the log linear phase of amplification instead of measurement at the plateau phase an accurate quantification of target molecules is possible. PCR efficiencies are taken in account. Specific detection is provided by use of fluorescence labeled hybridization probes. The sensitivity of our real-time PCR protocol was 1 AML1-ETO positive cell in the background of 100,000 normal cells. Relative quantification using standard curve analysis or an external calibrator sample revealed virtually identical results. Real-time PCR offers the opportunity to monitor the decline of AML1-ETO fusion transcripts during induction chemotherapy. Whether the initial decline or the detectable level of MRD after induction therapy is of prognostic relevance remains to be clarified, but has already been shown in acute lymphoblastic leukemia (ALL) [12]. Furthermore we demonstrated that G6PDH as well as cABL are appropriate housekeeping genes for relative quantification of AML1-ETO fusion transcripts. As real-time PCR is far more sensitive than cytomorphology, cytogenetics and FISH it has the potential to diagnose molecular relapse before clinical manifestation and therefore enable early clinical intervention. We conclude that quantitative real-time PCR using hybridization probes is a potent tool for monitoring minimal residual disease in t(8;21) positive AML during and after therapy. Standardization of sample preparation and PCR protocols is necessary to provide comparable results especially between different laboratories. It is likely that real-time PCR will become a widely used method in MRD detection of multiple diseases characterized by molecular markers. However, in the case of rapid proliferating diseases such as AML, frequent analysis of blood sample after therapy is necessary to detect molecular relapse before cytogenetic or hematologic relapse. This is especially important in relapsed AML because survival rates are less than 20% [2]. Further studies on MRD are needed in order to establish a biological basis on which clinical decisions are made and finally, most importantly, to improve overall survival.

References

1. Munker R, Hiller E, Paquette R (2000) Modern Hematology. Humana Press, Totowa, New Jersey
2. Loewenberg B, Downing JR, Burnett A (1999) Acute myeloid Leukemia. N Engl J Med 341:1051–1063
3. Bitter MA, Le Beau MM, Rowley JD, Larson RA, Golomb HM, Vardiman JW (1987) Associations between morphology, karyotype, and clinical features in myeloid leukemias. Hum Pathol 18(3):211–25
4. Liu Yin J A, Tobal K (1999) Detection of Minimal Residual Disease in Acute Myeloid Leukemia: Methodologies, Clonical and Biological Significance. Br J Haematol 106:578–590
5. Rowley JD (1998) The critical role of chromosome translocations in human leukemias. Annu Rev Genet. 32:495–519. Review
6. Downing JR, Head DR, Curcio-Brint AM, Hulshof MG, Motroni TA, Raimondi SC, Carroll AJ, Drabkin HA, Willman C, Theil KS, et al (1993) An AML1/ETO fusion transcript is consistently detected by RNA-based polymerase chain reaction in acute myelogenous leukemia containing the (8;21)(q22;q22) translocation. Blood 1;81(11):2860–5
7. Grimwade D, Walker H, Oliver F, Wheatley K, Harrison C, Harrison G, Rees J, Hann I, Stevens R, Burnett A, Goldstone A. (1998) The importance of diagnostic cytogenetics on outcome in AML: analysis of 1,612 patients entered into the MRC AML 10 trial. The Medical Research Council Adult and Children's Leukaemia Working Parties. Blood 1;92(7):2322–33
8. Asou H, Tashiro S, Hamamoto K, Otsuji A, Kita K, Kamada N (1991) Establishment of a human acute myeloid leukemia cell line (Kasumi-1) with 8;21 chromosome translocation. Blood 77(9):2031–6
9. Wittwer CT, Herrmann MG, Moss AA, Rasmussen RP (1997) Continuous fluorescence monitoring of rapid cycle DNA amplification. Biotechniques 22(1):130–1, 134–138
10. Wittwer CT, Ririe KM, Andrew RV, David DA Gundry RA Balis UJ The LightCycler: Amicrovolume multisample fluorimeter with rapid temperature control. Biotechniques 1997;22: 176
11. Lion T (1996) Appropriate controls for RT-PCR. Leukemia 10(11):1843. Review
12. Emig M, Saussele S, Wittor H, Weisser A, Reiter A, Willer A, Berger U, Hehlmann R, Cross NCP, Hochhaus A (1999) Accurate and Rapid Analysis of Residual Disease in Patients with CML Using Specific Fluorescent Hybridization Probes for Real Time Quantitative RT-PCR. Leukemia 13:1825–1832
13. Pui CH, Campana D (2000) New Definition of Remission in Childhood Acute Lymphoblastic Leukemia. Leukemia 14:783–785

Rapid Quantitative Detection of Free Cancer Cells in the Peritoneal Cavity of Gastric Cancer Patients with Real-Time CEA RT-PCR Using Hybridization Probes

Hayao Nakanishi*, Yasuhiro Kodera, Masae Tatematsu

Introduction

Detection of free cancer cells in the abdominal cavity has important prognostic implications for cancer of the stomach, pancreas, and ovary. Cytology examination of peritoneal washes has been the gold standard, and its clinical relevance is now recognized worldwide. The conventional method, however, is reported to lack sensitivity and patients with negative cytology results may nevertheless present with recurrent peritonitis carcinomatosa.

The recently introduced reverse-transcriptase polymerase chain reaction (RT-PCR) has demonstrated sensitive detection of micrometastases in the peritoneal cavity [1, 2]. Since carcinoembryonic antigen (CEA) is a specific marker for epithelial cells in peritoneal washes and the expression at the mRNA level is maintained in the majority of gastric cancers, it is recognized as a reliable target gene for detection of tumor cells with RT-PCR [3]. Despite such distinct advantages in terms of sensitivity, however, the RT-PCR analysis has the following shortcomings: (1) it is only qualitative and lacks quantitative assessment of risk of peritoneal recurrence; (2) a certain proportion of false-positive RT-PCR results appears to have been generated with a subsequent uneventful clinical course; and (3) gene amplification and subsequent data analysis are time-consuming and consequently the results of RT-PCR are not available during surgery.

To overcome these problems, we developed a rapid and quantitative method for detection of free cancer cells in peritoneal washes using a new generation of thermal cycler (the LightCycler system) with a hybridization probe format [4, 5]. This rapid cycle real-time PCR system allows quantification of the initial template copy number based on the inverse correlation between the cycle number at which the sample fluorescence exceeds the background level and the starting copy number. Consequently, real-time RT-PCR with the LightCycler instrument was found to be a sensitive, quantitative, specific and rapid method for detecting free cancer cells in peritoneal washes of patients with gastric cancer and has great promise for routine use in the clinical setting.

---

* Hayao Nakanishi (✉) (e-mail: hnakanis@aichi-cc.jp)
  Division of Oncological Pathology, Aichi Cancer Center Research Institute, Chikusa-ku, Nagoya 464–8681, Japan

Materials

Equipment LightCycler instrument

Reagents Isogen RNA extraction buffer (Nippon Gene, Tokyo, Japan)
Glycogen (Roche Diagnostics, Mannheim, Germany)
Random hexanucleotide primer (Pharmacia Biotech, Uppsala, Sweden)
SuperScript II RNase H-reverse transcriptase (Gibco BRL, Gaithersburg, USA)
Amplification primer (Nihon Gene Research Laboratories, Sendai, Japan)
Hybridization probes (Nihon Gene Research Laboratories)
LightCycler-DNA Master hybridization probes (Roche Diagnostics)
TaqStart antibody (Clontech, Palo Alto, USA)

Procedure

Sample preparation Peritoneal washes (50–100 ml) obtained from the Douglas cavity at the beginning of each operation were centrifuged at 1,800 rpm for 5 min to collect intact cells, rinsed with PBS and dissolved in 1 ml Isogen. Total RNA was extracted using a guanidinium-isothiocyanate-phenol-chloroform-based method in the presence of glycogen (20 µg/sample). Extracted total RNA up to 5 µg were preincubated with 50 ng of random hexanucleotide primer in 9 µl of solution for 10 min at 70°C. After chilling on ice, the following were added: 4 µl of fivefold synthesis buffer (250 mM Tris-HCl, pH 8.3, 375 mM KCl, 15 mM MgCl2), 2 µl of 100 mM dithiothreitol, 4 µl of 2.5 mM of each dNTP and 1 µl of SuperScript II RNase H⁻reverse transcriptase (200 units/µl). The reaction mixture was preincubated for 10 min at 25°C and then incubated for 50 min at 42°C, followed by heating at 70°C for 15 min. The resultant first-strand cDNA was used for PCR amplification with the LightCycler instrument.

Oligonucleotides The CEA-specific oligonucleotide primers and hybridization probes used are designed with the aid of Oligo primer analysis software based on published sequences [5] and synthesized by Nihon Gene Research Laboratories. Details for primers and probes used in this study are summarized in Table 1. To quantify and

Table 1. Oligonucleotides

Carcinoembryonic antigen (GenBank Accession #M29540)				
Sequence	5′-Position	Length	GC (%)	T_m (°C)
Primers				
AACTTCTCCTGGTCTCTCAGCT	2148	22	50	63.9
GCAAATGCTTTAAGGAAGAAG	2272	21	38	61.9
Product	2148–2272	125 bp		
Hybridization probes				
TGAAATGAAGAAACTACACCAGG-F	2228	23	39	63.9
LCRed640-CTGCTATATCAGAGCAACCCCAA-P	2204	23	48	68.6

prove the integrity of the isolated RNA, a real-time RT-PCR analysis for glyceraldehyde-3-phosphate dehydrogenase (GAPDH) was also carried out using primers and hybridization probes [5].

PCR amplification was carried out with the following reaction mixture:

LightCycler PCR

	Volume [µl]	[Final]
LightCycler-DNA Master Hybridization Probes	2.0	1X
TaqStart antibody	0.2	
MgCl2 (25 mM)	2.4	4.0 mM
CEA primers (10 µM each)	1.0+1.0	0.5 each µM
Fluorescein-labeled probe (4.0 µM)	1.0	0.20 µM
LCRed640-labeled probe (8.0 µM)	1.0	0.40 µM
H$_2$O (PCR grade)	9.4	
Total volume	18.0	

0.2 µl of TaqStart antibody is added to each 2 µl of DNA Master Hybridization probes and pre-incubated for 5 min at room temperature before the addition of other reagents. 18 µl of master mixture and 2 µl cDNA were added to each capillary. Sealed capillaries were centrifuged (1000 g for 15 s) and placed in the LightCycler instrument.

The following PCR protocol was used for amplification and quantification:

- Denaturation for 90 sec at 95°C
- Amplification

Cycling program:

Parameter	Value		
Cycles	45		
Type	Quantification		
	Segment 1	Segment 2	Segment 3
Target temperature [°C]	95	50	72
Incubation time [s]	0	10	10
Temperature transition rate [°C/s]	20	20	20
Acquisition mode	None	Single	None
Gains	F1=1; F2=15		

External standards for CEA mRNA were prepared by tenfold serial dilutions (1–10^5 cells) of cDNA equivalent to 1×10^6 COLM-2 cells spiked to 1×10^7 peripheral blood leukocytes. External standards for GAPDH mRNA were also prepared by tenfold serial dilutions (10^2–10^7 cells) of cDNA equivalent to 1×10^7 peripheral blood leukocytes. Each run consisted of six external standards, a negative control without a template and patient samples with unknown mRNA concentrations. Quantitation of mRNA in each sample was then performed automatically by reference to the standard curve constructed each time according to the LightCycler software.

Results

Fig. 1 illustrates a standard curve constructed by plotting the log of tenfold serially diluted CEA expressing colon carcinoma cells (COLM-2) against the respective crossover point (Cp). The slope was –3.66 and the mean squared error γ was less than 0.5, indicating a good amplification efficiency (>90%) and a precise log-linear relation in the range of mRNA corresponding to $1–10^5$ carcinoma cells, reflecting a comparable sensitivity to conventional nested-primer RT-PCR with a wide measuring range. CEA mRNA values for patient samples with unknown concentration were calculated with reference to this calibration curve and representative examples are shown in Fig. 1. No CEA mRNA at significant levels was detected in peripheral blood from 15 healthy volunteers and primary human cultured mesothelial cells from the omentum, both being major cellular constituents of the peritoneal washes (data not shown).

Total assay time including pretreatment of the sample (15 min), cDNA synthesis from extracted RNA (90 min) and amplification and subsequent data analysis (45 min) was approximately 3 h, which is less than half the time required for conventional RT-PCR.

The mean CEA mRNA values for 37 negative control patients consisting of 4 patients with benign disease and 33 with mucosa-confined gastric cancer, considered to be clinically benign, was 0.64±2.54 (SD), ranging from 0 to 10.0. On the other hand, CEA mRNA values in the peritoneal washes from the Douglas cavity of patients with synchronous peritoneal metastases ranged from 17.1 to 41,920, except for two false-negative cases. Based on the control patient levels calculated from the formula (mean+SD ×2) and the lowest value in patients with peritoneal metastasis, we determined a cut-off level at 10 in the present study. Any samples with a CEA mRNA value less than 10 was classified as negative.

Quantitative analysis by the LightCycler instrument demonstrated average CEA mRNA values in peritoneal washes from the Douglas cavity ranging from 0.84 for pT1 tumors, 513 for pT2 tumors, 3178 for pT3 tumors and 3459 for pT4 tumors. These values correlated statistically with the depth of cancer invasion ($P<0.01$) (Fig. 2).

The quantitative real-time RT-PCR was compared to cytology and conventional RT-PCR in peritoneal washes from 109 gastric cancer patients. In patients belonging to pT1, pT2, pT3, and pT4 categories, the cytology examination detected free cancer cells in 2%, 5%, 41%, and 8%, and conventional RT-PCR gave positive results in 10%, 30%, 78%, and 69%, respectively. Real-time RT-PCR with the cut off value detected 4%, 15%, 67%, and 54%, respectively, giving intermediate detection rates between cytology and conventional RT-PCR. This suggests that false-positive results found in conventional RT-PCR may be eliminated with LightCycler PCR.

Moreover, CEA mRNA values for peritoneal washes in synchronous peritoneal metastasis-positive patients was more than 50-fold higher than in those without metastasis (data not shown). These results suggest a positive correlation between CEA mRNA levels in peritoneal washes and prognosis [5].

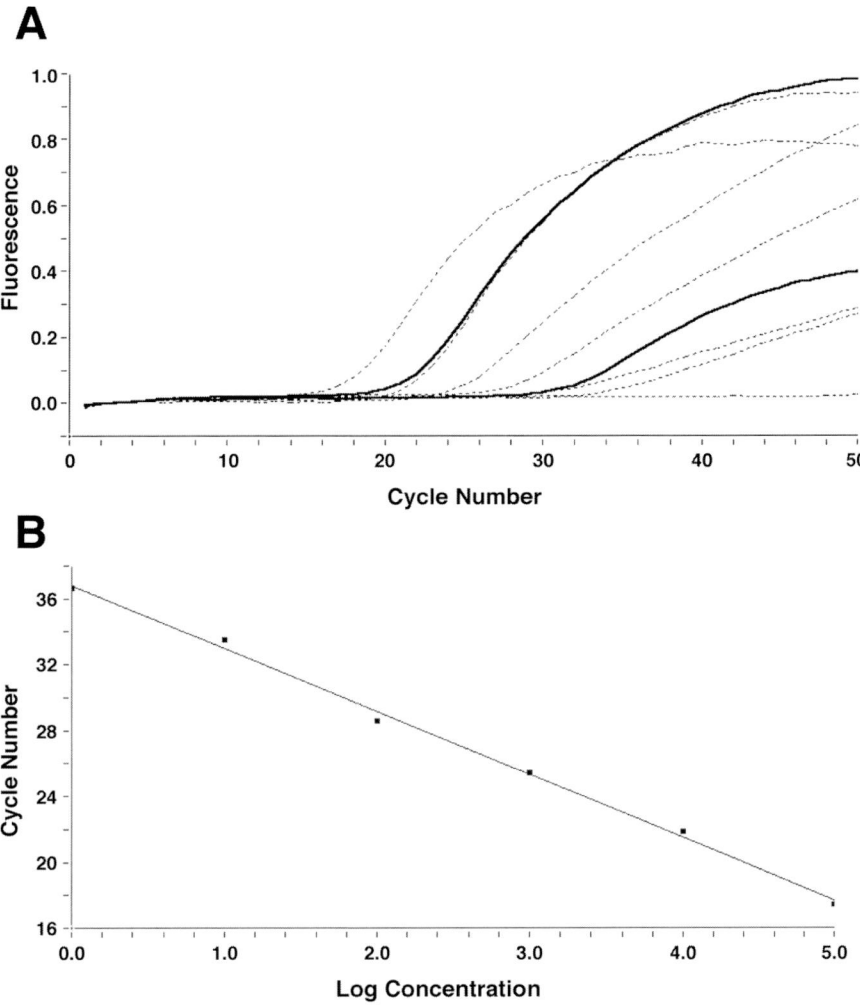

Fig. 1a,b. Representative results of real-time RT-PCR with the LightCycler. **a** Run profile of fluorescence vs PCR cycle. Six external standards (1–10^5 COLM-2 colon carcinoma cells equivalent cDNA, *dashed lines*) were compared with two patient peritoneal wash samples (*solid lines*) with unknown concentrations. *Flat dashed line*, no template control. Fifty rounds of amplification were completed within 45 min from the start. **b** A calibration curve for CEA mRNA estimation, constructed from data for six external controls shown in **a** by plotting the crossing point (Cp) against the log (COLM-2 cell number). Slope, –3.66; mean squared error, $f Á$ = 0.48. Relative CEA mRNA values in the patient samples were calculated with reference to this curve

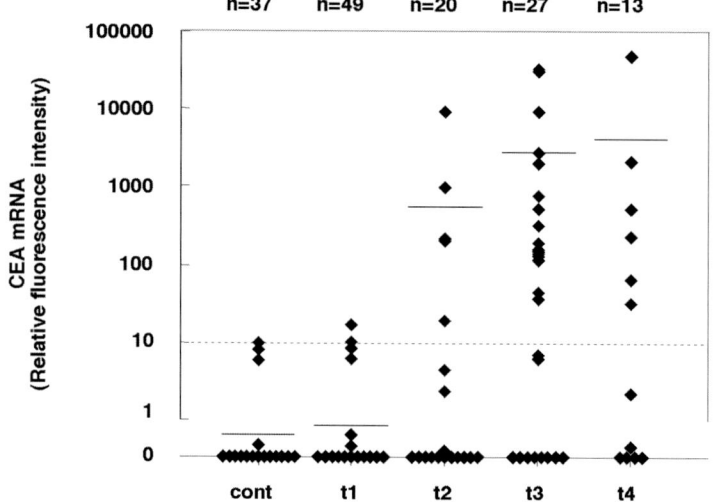

Fig. 2. Relative CEA mRNA values for peritoneal washes from the Douglas cavity measured by real-time RT-PCR with LightCycler in 109 patients with gastric cancer according to the depth of invasion (pT category). CEA mRNA values correlate statistically with the depth of cancer invasion ($P<0.01$). *Solid lines* indicate cut-off values. *From left to right* are controls, pT1 tumors with mucosal to submucosal invasion, pT2 tumors with muscularis propria to subserosal invasion, pT3 tumors with serosal invasion, and pT4 tumors with invasion to adjacent tissues

Comments

For normalization of CEA mRNA values, we also quantified GAPDH mRNA as an internal control by the LightCycler instrument using the hybridization probe format. The CEA/GAPDH ratio correlated well with the depth of tumor invasion, like the non-normalized data, which suggests that there is no need for correction of CEA mRNA values with reference to GAPDH mRNA. Since the total number of cancer cells rather than the cancer cell/noncancer cell ratio in peritoneal washes may be the hallmark of progression of metastatic spread, we used uncorrected CEA mRNA rather than the CEA/GAPDH ratio for a more convenient and appropriate parameter to monitor peritoneal recurrence in this study.

The key to successful quantification, especially in the low copy range, lies in the storage of the external standards. We prefer to prepare multiple aliquots, in order to avoid repetitive freezing and thawing. We dilute them to our working concentration and store them in aliquots at −30°C. For long-term storage, −80°C is optimal.

References

1. Nakanishi H, Kodera Y, Torii A, Hirai T, Yamamura Y, Kato T, Kito T, Tatematsu M (1997) Detection of carcinoembryonic antigen-expressing free tumor cells in peritoneal washes from patients with gastric carcinoma by polymerase chain reaction. Jpn J Cancer Res 88: 687–692
2. Kodera Y, Nakanishi H, Yamamura Y, Shimizu Y, Torii A, Hirai T, Yasui K, Morimoto T, Kato T, Kito T, Tatematsu M (1998) Prognostic value and clinical implications of disseminated cancer cells in the peritoneal cavity detected by reverse transcriptase-polymerase chain reaction and cytology. Int J Cancer 79:429–433
3. Gerhard M, Juhl H, Kalthoff H, Schreiber HW, Wagener C, Neumaier M (1994) Specific detection of carcinoembryonic antigen-expressing tumor cells in bone marrow aspirates by polymerase chain reaction. J Clin Oncol 12:725–729
4. Nakanishi H, Kodera Y, Yamamura Y, Kuzuya K, Nakanishi T, Ezaki T, Tatematsu M (1999) Molecular diagnostic detection of free cancer cells in the peritoneal cavity of patients with gastrointestinal and gynecologic malignancies. Cancer Chemother Pharmacol 43 [Suppl]S32–S36
5. Nakanishi H, Kodera Y, Yamamura Y, Ito S, Kato T, Ezaki T, Tatematsu M (2000) Rapid quantitative detection of carcinoembryonic antigen-expressing free tumor cells in the peritoneal cavity of gastric cancer patients with real-time RT-PCR on the LightCycler. Int J Cancer 89:411–417

Quantitative Measurement of the mRNA Expression of the Tumor-Associated Antigen PRAME by Real-Time RT-PCR Using LightCycler and SYBR Green I Technology

Jochen Greiner, Mark Ringhoffer, Anita Szmaragowska, Sandra Hübsch, Hartmut Döhner, Michael Schmitt*

Introduction

Characterization of immunogenic tumor-associated antigens (TAA) is mandatory for the design of cancer vaccines. Cancer antigens eliciting an immune response in the tumor-bearing host are targets for specific immunotherapy. In recent years, many TAA have been identified by methods using cytotoxic T-lymphocytes or by serological screening of recombinant expression cloning. TAA can be classified into three major categories:
1. Overexpressed genes are known to elicit immune responses by overriding thresholds critical for the maintenance of tolerance [1]. These antigens are not strictly tumor-specific, but they are overexpressed in different tumor tissues.
2. Antigens encoded by mutated genes induce immune responses by changing the expression and/or conformation of proteins [2].
3. Cancer/testis (CT) antigens are a family of tumor genes expressed exclusively in cancer cells and in testis tissue [3].

The TAA PRAME (Preferentially expressed antigen of melanoma) is such an antigen with high expression in tumor and testis. It is also expressed at a low level in the endometrium and ovary [4]. PRAME has been characterized in a melanoma cell line derived from a patient (LB33) [4]. The cellular function of PRAME remains to be elucidated. PRAME is frequently expressed in melanoma (91%), lung squamous cell carcinoma (78%), sarcoma (39%), and renal cell cancer (41%) [4, 5]. It is also expressed in hematological malignancies such as acute and chronic myeloid leukemias and in different lymphoma subtypes [6, 7]. Approximately 47% of acute myeloid leukemia (AML) samples obtained from patients diagnosed at our institution expressed PRAME at a high mRNA level [7], while no mRNA expression was found in peripheral blood, bone-marrow cells and CD34-positive separated cell samples of healthy volunteers by RT-PCR [6, 7]. The fact that PRAME is not expressed in most other normal tissues [4, 6] reduces the risk for negative effects of a vaccination with the antigen PRAME. It is therefore an inter-

---

* Michael Schmitt (✉) (e-mail: michael.schmitt@medizin.uni-ulm.de)
  Jochen Greiner, Mark Ringhoffer, Anita Szmaragowska, Sandra Hübsch, Hartmut Döhner
  Third Department of Medicine, University of Ulm, 89070 Ulm, Germany

esting target for tumor immunotherapy. Clinical trials must however include a precise monitoring of possible adverse effects on tissue with very low expression of PRAME. Moreover, PRAME might be a possible marker for minimal residual disease. Real-time PCR using the LightCycler method allows an accurate quantification of the mRNA expression level of the TAA PRAME in tumors and peripheral blood and bone-marrow samples. Detecting the mRNA expression of this antigen during the course of chemotherapy and/or hematopoietic transplantation might constitute a useful tool for the monitoring of minimal residual disease in AML. Moreover, real-time RT-PCR might be helpful to select potential candidates for a PRAME specific immunotherapy among AML patients.

Materials

Equipment
LightCycler instrument (Roche Diagnostics, Mannheim, Germany)
LightCycler software vers. 3.39 (Roche Diagnostics)
LightCycler Capillaries (Roche Diagnostics)
Mac DNASISPro version 3.5 Primer design software (Hitachi)

Reagents
mRNA Quickprep Micro Purification Kit (Amersham Pharmacia Biotech, Little Chalfont, England, UK)
cDNA Preparation Kit (Superscript II Gibco BRL, Frederick, MA)
Amplification Primers (MWG-Biotech, Munich, Germany)
LightCycler FastStart Kit (Roche Diagnostics)
NuSieve agarose (FMC Bio Products)
Qiaquick PCR Purification Kit (Chatsworth, CA, USA)

Procedure

Cell Culture
Cells of the human suspension cell line K562 (a CML cell line established from a patient with chronic myeloid leukemia in blast crisis) expressed PRAME at a high level. The cell line was cultured in RPMI 1640 (Biochrom, Berlin, Germany) containing 10% (v/v) fetal calf serum, 2 mM L-glutamine, 100 units/ml penicillin and 100 units/ml streptomycin.

Sample Preparation
Total RNA was isolated by the phenol/chloroform method (Chomczynski and Sacchi [8]). The mRNA was prepared using the mRNA QuickPrep Micro purification kit (Amersham Pharmacia Biotech, Little Chalfont, England, UK). In total, 2.5 µg mRNA of each sample was transcribed to cDNA (Superscript II Gibco BRL, Frederick, Maryland). For the quantification of the antigen PRAME, the mRNA expression of PRAME was correlated with the mRNA expression of the house-keeping genes TBP (TATA-binding protein) and β-actin. The primers were used to amplify an 833-bp fragment of the TAA PRAME (Table 1) as described in the literature [4] and verified with DNASIS software. RT-PCR was performed under the described time conditions (Table 2), temperatures (Table 3), and with the indicated ingredients.

Table 1. Oligonucleotides

BLAST Search Accession # NM_006115, NM_003194 and XM_037235				
Position	Length		GC (%)	T_m (°C)
Primers-PRAME (NM_006115)				
GTCCTGAGGCCAGCCTAAGT	207	20	60	65.0
GGAGAGGAGGAGTCTACGCA	1039R	20	61	61.7
Primers-TBP (NM_003194)				
CACGAACCACGGCACTGATT	861	20	55	64.8
TTTTCTTGCTGCCAGTCTGGAC	949R	22	50	64.6
Primers-β-actin (XM_037235)				
GCATCGTGATGGACTCCG	522	18	61	61.9
GCTGGAAGGTGGACAGCGA	1133R	19	63	66.0

Table 2. LightCycler RT-PCR: parameters of amplification

	PRAME	β-actin	TBP
Initial Denaturation (min)	10	10	10
Denaturation (s)	5	15	15
Annealing (s)	5	10	10
Elongation (s)	40	25	20
Cycles	40	40	40

Table 3. LightCycler RT-PCR: temperatures for amplification

	PRAME	β-actin	TBP
Initial Denaturation (min)	94	95	95
Denaturation (s)	95	95	95
Annealing (s)	64	68	60
Elongation (s)	72	72	72

RT-PCR master mix:

	Volume [μl]	[Final]
LightCycler-DNA FastStart kit	2	1X
Primers (10 μM each)	1 (each)	0.5 μM
$MgCl_2$ (25 mM)	0.8	2.25 mM
Probes (see figures for concentration)	0.5	
Distilled water (PCR grade)	14.7	

- Melting Curve Analysis

Parameter	Value		
Cycles	1		
Type	Melting Curve		
	Segment 1	Segment 2	Segment 3
Target temperature [°C]	95	58	95
Incubation time [s]	0	10	0
Temperature transition rate [°C/s]	20	20	0.1
Acquisition mode	None	None	Cont.

Results

Expression Analysis of the Leukemia Antigen PRAME and of the House-Keeping Genes β-Actin and TBP by Real-Time PCR

To quantify the mRNA expression of PRAME, we established the real-time RT-PCR using the LightCycler SYBR Green I format. For the RT-PCR, $MgCl_2$ concentrations of 1.625, 2.0, 2.25 and 3 mM were evaluated. Best results were obtained with a concentration of 2.0 mM $MgCl_2$, so this concentration was used in further experiments. Figure 1a,b shows the sensitivity of the LightCycler RT-PCR detecting PRAME-positive leukemia cells in a sample of PRAME-negative mononuclear peripheral blood cells of a healthy volunteer. Normal mononuclear cells of peripheral blood showed no expression of PRAME. One PRAME-positive leukemia cell of the leukemia cell line K562 in 1×10^5 normal cells was detectable. Figure 1a shows the amplification and Fig. 1b the melting curve data of K562 cells diluted in normal mononuclear cells.

The quantitative expression of the antigen PRAME was evaluated using serial dilutions of a PRAME RT-PCR product. Figure 2a shows the amplification curves of different amounts of PRAME cDNA. The product was prepared by isolation (Qiaquick PCR Purification Kit, Qiagen) of a conventional RT-PCR product of PRAME. The concentration of the amplified DNA was measured by the photometer. The starting amounts of PRAME cDNA (11–0.11 pg) were used for preparation of standard curves. These curves were used as standard curves for the quantification of PRAME mRNA expression in cell lines and patients. The amplification of 50 ng of a K562 cDNA showed 2.5 pg of PRAME amplification starting product (sample 6). The melting curves of the quantification standard curves and of a K562 PCR product are shown in Fig. 2b. All melting curves showed a PRAME-specific peak at 90°C.

The mRNA expression of PRAME was correlated with the mRNA expression of the house-keeping genes β-actin and TBP (no retro-pseudogenes known [9]). The LightCycler RT-PCR of these genes was established. Figure 3 shows the amplification data of the house-keeping gene β-actin. Different $MgCl_2$ concentrations (1.625; 2.25; 3.0 and 4.0 mM) were tested for β-actin RT-PCR. For further experiments a $MgCl_2$ concentration of 2.25 mM was used. Figure 4a shows the amplification curves of serial diluted cDNA samples of the house-keeping gene TBP. These samples were prepared by isolation of a conventional RT-PCR product of

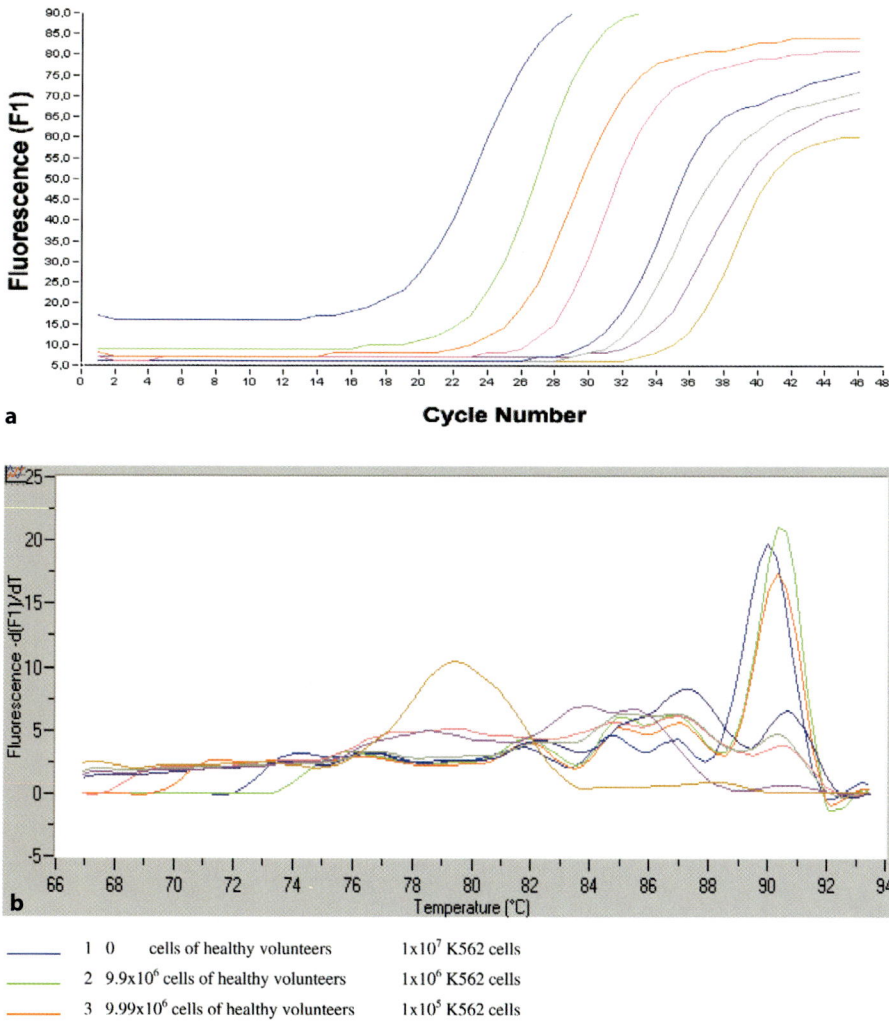

—	1 0 cells of healthy volunteers	1×10^7 K562 cells
—	2 9.9×10^6 cells of healthy volunteers	1×10^6 K562 cells
—	3 9.99×10^6 cells of healthy volunteers	1×10^5 K562 cells
—	4 9.999×10^6 cells of healthy volunteers	1×10^4 K562 cells
—	5 1×10^7 cells of healthy volunteers	1×10^3 K562 cells
—	6 1×10^7 cells of healthy volunteers	1×10^2 K562 cells
—	7 1×10^7 cells of healthy volunteers	1×10^1 K562 cells
—	8 1×10^7 cells of healthy volunteers	1×10^0 K562 cells

Fig. 1a,b. Sensitivity of the PRAME LightCycler RT-PCR detecting leukemia K562 cells (PRAME positive) mixed with normal mononuclear cells (PRAME negative). **a** Amplification curves, **b** melting curve analysis. With the LightCycler RT-PCR a ratio of one leukemia cell (positive for PRAME) in 1×10^5 normal mononuclear cells (negative for PRAME) could be detected

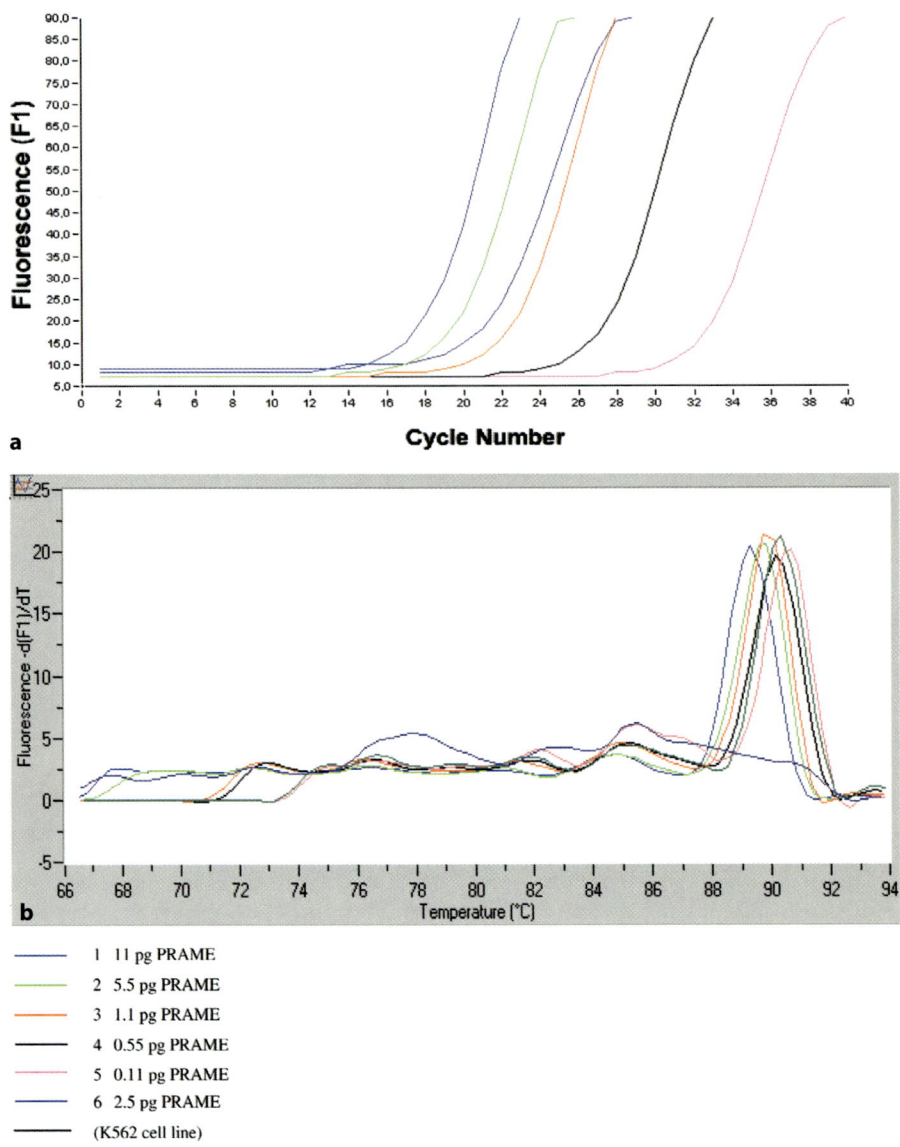

    1 11 pg PRAME
    2 5.5 pg PRAME
    3 1.1 pg PRAME
    4 0.55 pg PRAME
    5 0.11 pg PRAME
    6 2.5 pg PRAME
       (K562 cell line)

Fig. 2a,b. Quantification of the antigen PRAME by serial dilution of a PRAME-PCR product. The amplification curves of different amounts of PRAME cDNA (**a**) and the melting curves of the standard curves (**b**). PCR of 50 ng K562 cDNA showed 2.5 pg of PRAME amplification starting product

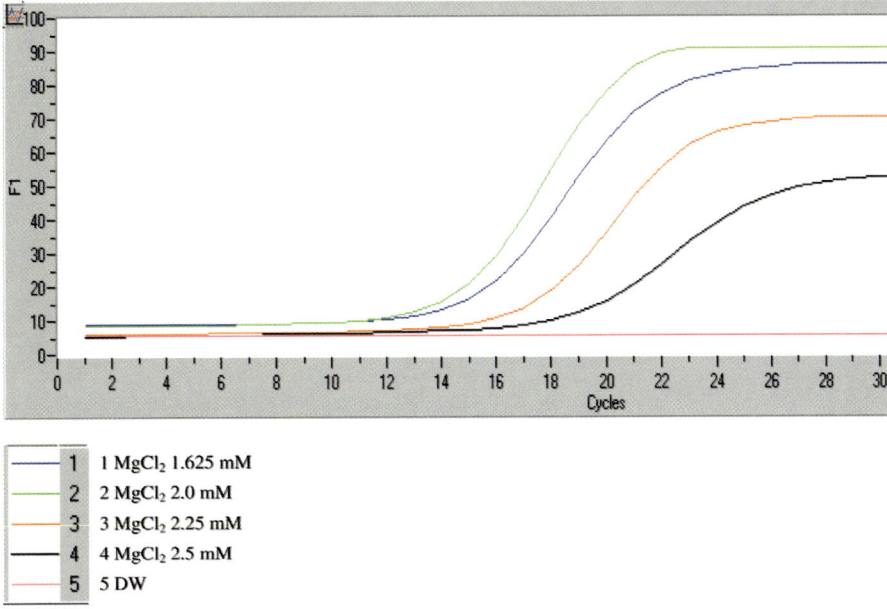

Fig. 3. Real-time RT-PCR using the LightCycler method of the house-keeping gene β-actin. Amplification curve of K562 cDNA RT-PCR with different MgCl$_2$ concentrations

TBP. The concentration of the amplified DNA was measured by the photometer. These results were used as standard curves for quantification of TBP mRNA expression. The regression analysis of these curves ($r=0.98$) is shown in Fig. 4b. The mRNA expression of PRAME was correlated with the mRNA expression of TBP or β-actin as follows:

$$c[\text{PRAME}] = \frac{m(\text{PRAME})}{m(\text{PRAMEstandard})} \times \frac{m(\text{TBPstandard})}{m(\text{TBP})}$$

m = amount of cDNA
c = relative expression of PRAME

Comments

Immunotherapy alone or in combination with chemotherapy is a new form of cancer treatment. One possible approach in immunotherapy is the application of specific humanized monoclonal antibodies (e.g., anti-CD20 in B-cell lymphoma or anti-Her2/neu in breast cancer [10, 11]). Other forms of immunotherapy are cytokines like interleukin-2 and interferon for therapy of renal cell carcinoma, infusion of donor lymphocytes after stem cell transplantation of AML or CML

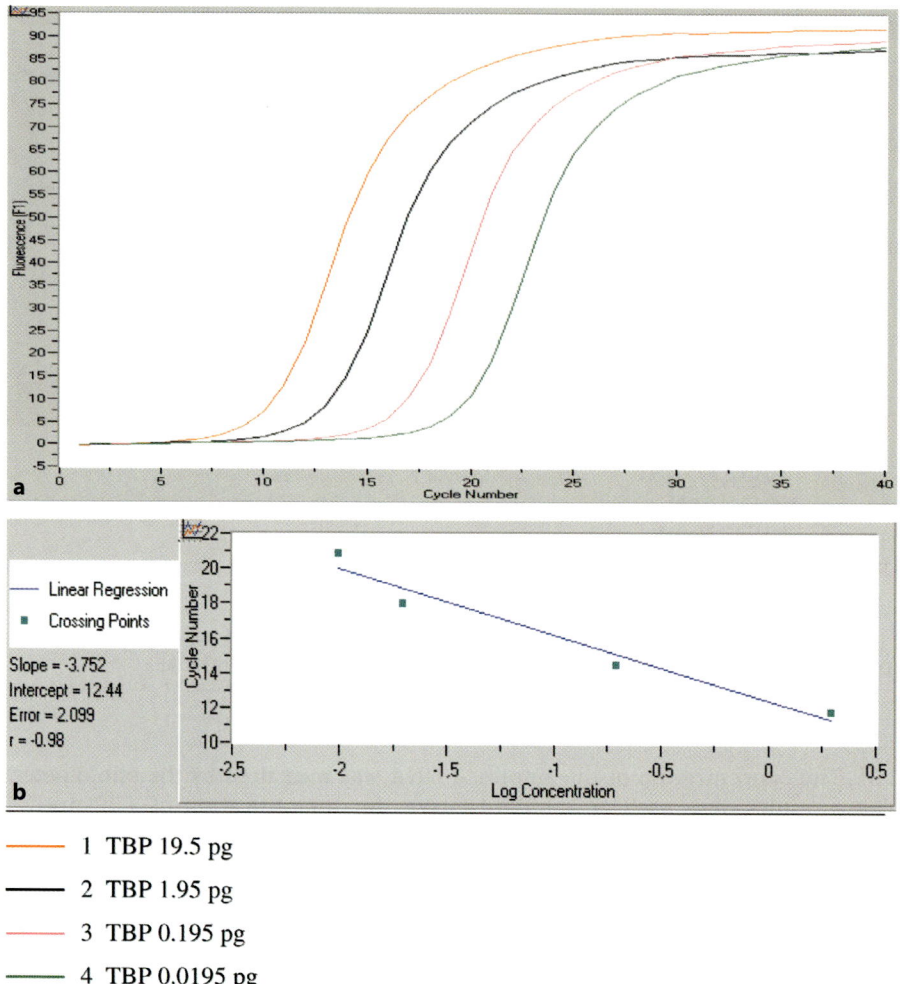

———— 1 TBP 19.5 pg
———— 2 TBP 1.95 pg
———— 3 TBP 0.195 pg
———— 4 TBP 0.0195 pg

Fig. 4a,b. Real-time RT-PCR using the LightCycler method of the house-keeping gene TBP. **a** The amplification curves of different amounts of the TBP RT-PCR product. These results were used as standard curves for the quantification of TBP mRNA expression. **b** The regression analysis of these curves ($r=0.98$)

patients or specific vaccination strategies with peptide-pulsed dendritic cells in melanoma. Efficient specific vaccination strategies depend on the definition of adequate target structures recognized by cytotoxic and helper T lymphocytes resulting in cellular or humoral immune responses. The tumor-associated antigen PRAME is expressed at a high level in different solid tumors, but also in hematological malignancies like acute myeloid leukemia. PRAME was the first tumor-associated cancer/testis antigen with high mRNA expression detected in leukemia

[4, 6]. PRAME-specific cytotoxic T lymphocytes (CTL) could be generated. The PRAME-specific CTL clone 17 (anti-LB33-E) recognizes a 9-mer peptide presented by HLA-A24 [4]. Recently, four different antigen epitopes of PRAME were characterized. They were presented on HLA-A0201 and induced specific CTL clones [12], thus indicating the immunogenic potential of PRAME. Therefore, leukemia patients with high mRNA expression of PRAME are potential candidates for PRAME-specific cellular immunotherapy. No mRNA expression was detected in peripheral mononuclear blood cells of healthy volunteers and in CD34-positive cells of early hematopoiesis of healthy volunteers [6, 7]. This characteristic of PRAME is of crucial importance for a PRAME-directed specific immunotherapy. Immunotherapy should not damage other tissues or suppress normal hematopoiesis in bone-marrow. Another potential target for immunotherapy in AML patients is the Wilms tumor antigen (Wt-1). Wt-1 is a gene expressed during embryonal cell differentiation and is correlated with a poor clinical prognosis in AML [13]. Wt-1 expression was described in 77% of AML patients but also in CD34-positive selected cell samples [14]. CD34+ separated cell samples also expressed Wt-1 at high or moderate levels, so immunotherapy using Wt-1 might suppress hematopoiesis in these cases. Therefore, PRAME is the most interesting target structure for immunotherapy in AML patients. Real-time RT-PCR using the LightCycler SYBR Green I technology allows accurate quantification of PRAME mRNA expression. The mRNA expression levels were correlated to the mRNA expression of the house-keeping genes TBP and β-actin. TBP was expressed at a more constant level than β-actin in samples of both healthy volunteers and tumor patients. In this study, real-time RT-PCR analysis showed a sensitivity of PRAME detection of one PRAME-positive AML cell in 1×10^5 normal PRAME-negative mononuclear cells. Minimal residual disease of PRAME-positive malignant cells of AML patients can be evaluated with LightCycler real-time RT-PCR during therapy like polychemotherapy or hematopoietic stem cell transplantation. In summary, PRAME may be a marker for minimal residual disease and a target for specific cellular immunotherapy in AML, providing both a diagnostic and a therapeutic tool.

References

1. Viola A, Lanzavecchia A (1996) T cell activation determined by T cell receptor number and tunable thresholds. Science 273:104–106
2. Coulie PG, Lehmann F, Lethe B, Hermann J, Lurquin C, Andrawiss M, Boon T (1995) A mutated intron sequence codes for an antigenic peptide recognized by cytolytic T lymphocytes on a human melanoma. Proc Natl Acad Sci 92:7976–7980
3. Chen YT, Scanlan MJ, Sahin U, Türeci Ö, Gure AO, Tsang S, Williamson B, Stockert E, Pfreundschuh M, Old LJ (1997) A testicular antigen aberrantly expressed in human cancers detected by autologous antibody screening. Proc Natl Acad Sci 94:1914–1918
4. Ikeda H, Lethe B, Lehmann F, Van Baren N, Baurain JF, De Smet C, Chambost H, Vitale M, Moretta A, Boon T, Coulie PG (1997) Characterization of an antigen that is recognized on a melanoma showing partial HLA loss by CTL expressing an NK inhibitory receptor. Immunity 6:199–208
5. Neumann E, Engelsberg A, Decker J, Störkel S, Jäger E, Huber C, Seliger B (1998) Heterogeneous expression of the tumor-associated antigens RAGE-1, PRAME, and glycoprotein 75 in

human renal cell carcinoma: candidates for T-cell-based immunotherapies. Cancer Res 58:4090–4095
6. Van Baren N, Chambost H, Ferrant A, Michaux L, Ikeda H, Millard I, Olive D, Boon T, Coulie PG (1998) PRAME, a gene encoding an antigen recognized on a human melanoma by cytolytic T cells, is expressed in acute leukaemia cells. Br J Haematol 102:1376–1379
7. Greiner J, Ringhoffer M, Simikopinko O, Szmaragowska A, Maurer U, Bergmann L, Schmitt M (2000) Simultaneous expression of different immunogenic antigens in acute myeloid leukemia (AML). Exp J Heamatol 28:1413–1422
8. Chomczynski P, Sacchi N (1987) Single-step method of RNA isolation by acid guanidinium thiocyanate-phenol-chloroform extraction. Anal Biochem 162:156
9. Bièche I, Laurendeau I, Tozlu S, Olivi M, Vidaud D, Lidereau R, Vidaud M (1999) Quantitation of MYC Gene Expression in Sporadic Breast Tumors with a Real-Time Reverse Transcription-PCR Assay. Cancer Res 59:2759–2765
10. Slamon DJ, Godolphin W, Jones LA, Holt JA, Wong SG, Keith DE, Levin WJ, Stuart SG, Udove J, Ullrich A, Press MF (1998) Studies of the HER-2/neu proto-oncogene in human breast and ovarian cancer. Science 244:707–712
11. Czuczman MS, Grillo-Lopez AJ, White CA, Saleh M, Gordon L, LoBuglio AF, Jonas C, Klippenstein D, Dallaire B, Varns C (1999) Treatment of patients with low-grade B-cell lymphoma with the combination of chimeric anti-CD20 monoclonal antibody and CHOP chemotherapy. J Clin Oncol 17:268–276
12. Kessler JH, Beekman NJ, Bres-Vloemans SA, Verdijk P, Veelen van PA, Kloostermann-Joosten AM, Vissers DCJ, Bosch GJA, Kester MGD, Sijts A, Drijthout JW, Ossendorp F, Offringa R, Melief CJM (2001) Efficient identification of novel HLA-A 0201-presented cytotoxic T lymphocyte epitopes in the widely expressed tumor antigen PRAME by proteasome-mediated digestion analysis. J Exp Med 193:73–88
13. Bergmann L, Miething C, Maurer U, Brieger J, Karakas T, Weidmann E, Hoelzer D (1997) High levels of Wilms' tumor gene (wt1) mRNA in acute myeloid leukemias are associated with a worse long-term outcome. Blood 90(3):1217–25
14. Maurer U, Weidmann E, Karakas T, Hoelzer D, Bergmann L (1997) Wilms tumor gene (wt1) mRNA is equally expressed in blast cells from acute myeloid leukemia and normal CD34+ progenitors. Blood 90:4230–4232

Expression Analysis of Telomerase-Genes hTERT and hTR by Quantitative PCR on LightCycler

BERND FRODERMANN, CHRISTOPHER POREMBA*

Introduction

Telomerase activity (TA) has been shown to be a strong indicator of cellular malignancy in virtually all malignant tumors [1–5]. Telomerase describes a ribonucleoprotein polymerase which uses an internal RNA component as a template to synthesize telomeric DNA directly onto the ends of chromosomes [6]. These repetitive sequences are considered to be important in the protection and replication of chromosomes. Cells without TA display progressive shortening of telomeric repeats with each cell division because lagging-strand DNA synthesis at the very end of linear chromosomes cannot be completed. Induction of TA in tumor cells could give rise to clonal immortality by compensating for the loss of telomeric DNA and thus maintaining telomere length. Recently, the catalytic subunit hTERT (or hTRT, hEST2, TP2) of human telomerase was cloned, and detection of hTERT expression by RT-PCR revealed a strong correlation with TA by the TRAP assay in the majority of tumors so far examined [7]. Other components of the telomerase holoenzyme complex, such as human telomerase RNA (hTR) and telomerase protein 1 (TP1, TLP1, hTEP1), seem to be expressed in both normal and tumor tissues, and expression levels of these genes revealed no or only limited correlation with TA [8, 9].

In this chapter, we describe two methods for quantitative real-time RT-PCR to analyze hTERT and hTR RNA expression on the LightCycler: the first method is based on the commercially available hTERT and hTR quantification kits from Roche Diagnostics, whereas the second method is based on our own protocol [10] that we developed and published before commercial kits became available.

Commercial Kit Protocol: Materials

LightCycler instrument (Roche Diagnostics, Mannheim, Germany) **Equipment**
LightCycler Capillaries (Roche Diagnostics)

---

* Christopher Poremba (✉) (e-mail: poremba@uni-muenster.de)
  Gerhard-Domagk-Institute of Pathology, Westfälische Wilhems-University, Domagkstrasse 17, 48149 Münster, Germany

Kits	LightCycler Telo TAGGGG hTERT Quantification Kit (Roche Diagnostics)
	LightCycler Telo TAGGGG hTR Quantification Kit (Roche Diagnostics)
Reagents	peqGOLD TriFast (peqLab, Erlangen, Germany)
	bovine pancreatic DNase I (Eurogentec, Seraing, Belgium)
	RNAguard Ribonuclease Inhibitor, Porcine (Amersham Pharmacia Biotech, Freiburg, Germany)

Procedure

Sample Preparation

RNA is isolated from fresh or frozen tissue samples (solid or cell culture) using peqGOLD TriFast in accordance with the manufacturer's protocol. The RNA is purified by DNase I digestion and resuspended in DEPC-treated water to obtain a final RNA concentration between 100 and 2,000 ng/µl, which is measured spectrophotometrically.

For analysis, the RNA is diluted to 50 ng/µl for hTR quantification and 50–100 ng/µl for hTERT quantification, as recommended in the manufacturer's protocol.

Oligonucleotide Design

Telomerase (hTERT)-encoding mRNA and telomerase-associated RNA (hTR) is reverse transcribed and a fragment of the generated cDNA is amplified with specific primers. The amplicon is detected by fluorescence using a specific pair of hybridization probes. The hybridization probes consist of two different short oligonucleotides that hybridize to an internal sequence of the amplified fragment during the annealing phase of the amplification cycle. One probe is labeled at the 5′-end with LightCycler Red 640, and to avoid extensions, modified at the 3′-end by phosphorylation. The other probe is labeled at the 3′-end with fluorescein.

The sequences of the primers and probes are not published.

The primers for hTERT and porphobilinogen desaminase (PBGD) are specific for mRNA because their sequences span exon splice junctions. Coamplification of contaminating genomic DNA is therefore prevented. However, as hTR is an intron-free gene, the primers also amplify genomic DNA. Because of this, controls without reverse transcriptase (RT) are strongly recommended, to make sure that there is no contaminating DNA amplified (see hTR Master Mix).

We have found that RNA purification by DNase I digestion eliminates the contaminating DNA sufficiently, so the reverse transcriptase negative samples did not produce a signal. Therefore, we do not currently perform these controls on a regular basis.

LightCycler PCR

Each Quantification Kit uses PBGD as a housekeeping gene, which is simultaneously quantified with the gene of interest in a "one-step RT-PCR".

For quantification, five standards of in-vitro transcribed hTERT or hTR RNA are provided with concentrations within the linear range of the reaction from 10^2 to 10^6 copies, or from 10^3 to 10^7 copies, respectively, per 2 µl. The exact number of copies is given by the manufacturer. Additional positive controls for hTERT or hTR, respectively, and PBGD are included.

As the procedure for each reaction is very similar and the reactions are optimized for the same conditions, the procedures for the hTERT and hTR kit are described together.

Depending on the gene to be quantified, two master mixes are needed: one for hTERT or hTR and one for PBGD. The setup for each run on the LightCycler instrument is as follows:
- Five capillaries for standards (hTR or hTERT, respectively)
- Four capillaries for hTR or hTERT and PBGD positive and negative controls
- Twice the number of samples (n), for hTR or hTERT and PBGD

- hTR: for detecting DNA contamination, prepare a third master mix without RT (-RT) for each sample. Three capillaries for each sample are then required.
- hTERT and hTR are the most interesting known telomerase genes, and one can use these kits to quantify both genes on one LightCycler run. This will save time and material, as the quantification of the housekeeping gene and the dilution of the sample is performed only once.

Notes

For the hTERT quantification kit, the hTERT master mix for each 20-μl reaction is as follows:

	Volume [μl]	[Final]
hTERT reaction mix	2	1×
Reverse transcriptase (RT)	0.1	
hTERT detection mix	2	1×
H$_2$O (PCR grade)	13.9	
Total volume	18	
Capillaries needed	7+n	

For the hTR quantification kit, the hTR master mix for each 20-μl reaction, with and without reverse transcriptase (RT) is as follows:

	Volume [μl]		[Final]
	+RT	−RT	
hTR reaction mix	2	2	1×
Reverse transcriptase (RT)	0.1		
hTR detection mix	2	2	1×
H$_2$O (PCR grade)	13.9	14	
Total volume	18	18	
Capillaries needed	7+n	n	

The housekeeping gene for both kits and the PBGD master mix for each 20-μl reaction:

	Volume [μl]	[Final]
hTERT/hTR reaction mix	2	1×
Reverse transcriptase (RT)	0.1	
PBGD detection mix	2	1×
H$_2$O (PCR grade)	13.9	
Total volume	18	
Capillaries needed	2+n	

A total of 18 μl of the appropriate master mix and 2 μl of the sample RNA (100 ng or 100–200 ng for hTERT or hTR, respectively) are added to each glass capillary placed in cooled adapters, using the following pattern:

Capillary	Reaction	hTERT/hTR master mix	PBGD master mix
1	Standard 1	1 reaction	
2	Standard 2	1 reaction	
3	Standard 3	1 reaction	
4	Standard 4	1 reaction	
5	Standard 5	1 reaction	
6	H$_2$O control	1 reaction	
7	Positive control	1 reaction	
8	H$_2$O control		1 reaction
9	Positive control		1 reaction
10–32	Samples of interest (n)	n reactions	n reactions
	Number of reactions	7+n	2+n

The sealed capillaries are briefly centrifuged (10 s at 700 g) and put into the Light-Cycler rotor.

The following RT-PCR protocol is used:
- Reverse transcription at 60°C for 10 min
- Denaturation at 95°C for 30 s
- Amplification

Parameter	Value		
Cycles	45 for hTR; 40 for hTERT		
Type	Quantification		
	Segment 1	Segment 2	Segment 3
Target temperature [°C]	95	60	72
Incubation time [s]	0	10	10
Temperature transition rate [°C/s]	20	20	20
Acquisition mode	None	Single	None
Gains	F1=1; F2=15; F3=30		

- Cooling for 1 min at 40°C
  Set "display mode" to fluorescence channel 2/1 (F 2/1).

Results

The fluorescence profiles and quantification of hTERT and hTR from one sample, the five standards and positive and negative controls are shown in Figs. 1 and 2.

For analysis of the data, the F2/F1 display mode was used. The quantities of the transcripts are calculated by using the "second derivative maximum" method. We have found a better reproducibility and a lower user dependence than using the "fit-point" method.

Error and Correlation

We accepted only a correlation coefficient (r) of −1.00, to keep a high validity of our results.

The relative expression level is calculated by dividing the amount of hTERT or hTR transcripts, respectively, by the amount of PBGD transcripts.

We recommend repeating the quantification at least once to validate the results.

In-House Protocol: Materials

Equipment

LightCycler Instrument (Roche Diagnostics)
LightCycler Capillaries (Roche Diagnostics)

Kits

LightCycler-DNA Master SYBR Green I (Roche Diagnostics)
First-Strand cDNA Synthesis Kit (Amersham Pharmacia Biotech)

Reagents

peqGOLD TriFast (peqLab)
bovine pancreatic DNase I (Eurogentec)
RNAguard Ribonuclease Inhibitor, Porcine (Amersham Pharmacia Biotech)
TaqStart Antibody (Clontech, Heidelberg, Germany)
Amplification primers (Amersham Pharmacia Biotech)
Telomerase positive tissue sample or cell line, e.g., Ewing tumor cell line VH64

Procedure

Sample Preparation

For RNA isolation and purification of the tissue samples of interest and of the telomerase positive reference sample, see "Sample Preparation" under "Commercial Kit Protocol".

From the samples, cDNA is synthesized from approximately 2 µg of RNA using the First-Strand cDNA Synthesis Kit from Amersham Pharmacia Biotech and stored at −20°C.

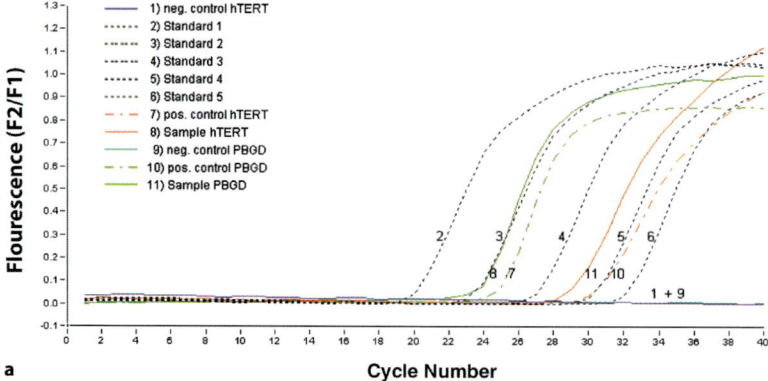

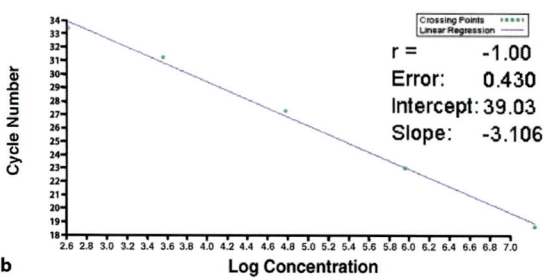

hTERT - Calculation:	Calculated Concentration	Crossingpoint (Cycle number)
1) neg. control hTERT	-	-
2) Standard 1 (1.2×10^6 copies)	1,388,000	19.95
3) Standard 2 (9.5×10^4 copies)	96,960	23.54
4) Standard 3 (9.2×10^3 copies)	7,246	27.04
5) Standard 4 (8.3×10^2 copies)	651.5	30.29
6) Standard 5 (1.3×10^2 copies)	178.1	32.04
7) pos. control hTERT	566.1	30.48
8) Sample hTERT	1,736	28.97
9) neg. control PBGD	-	-
10) pos. control PBGD	42,590	24.65
11) Sample PBGD	84,540	23.73

$$rel.\,hTERT\;expression = \frac{Sample\,hTERT}{Sample\,PBGD} = \frac{1,736}{84,540} = 2.1*10^{-2}$$

Fig. 1. LightCycler run for hTERT quantification. **a** Fluorescence using channel F2 / F1 is plotted against PCR cycle number. hTERT and PBGD expression levels were determined for the commercial kit positive control (capillaries 7 and 10) and a sample of interest (capillaries 8 and 11). Five samples with known hTERT copy-numbers were provided in the Kit and included as standards (capillaries 2–6). **b** Linear regression of the 5 standards was determined by using the 'Second Derivative Maximum' method. The linear regression and calculation of the correlation coefficient, r, serves to confirm accuracy and reproducibility. **c** The relative hTERT expression in the Sample was determined as a ratio of hTERT and PBGD levels which had been calculated from the standard curve

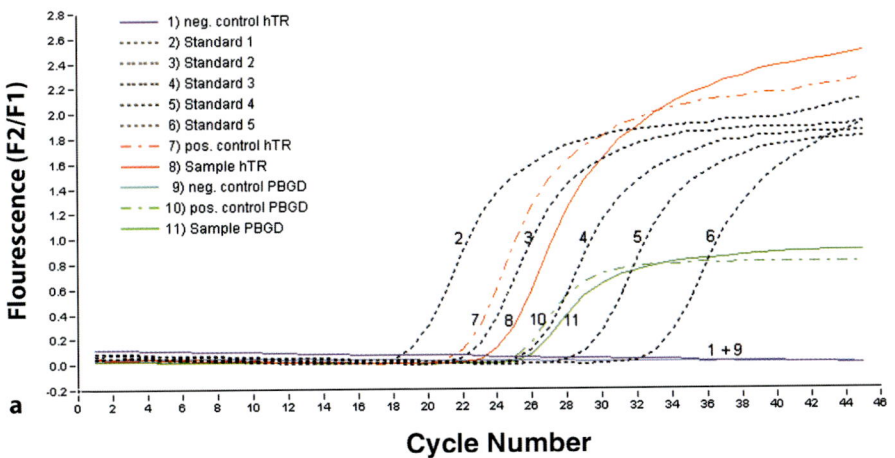

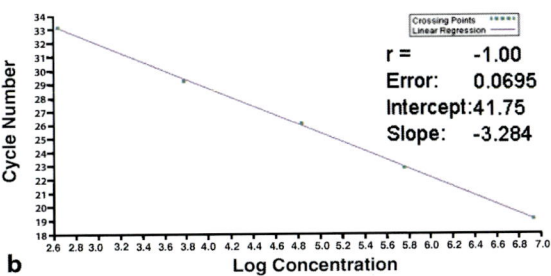

hTR - Calculation:	Calculated Concentration	Crossingpoint (Cycle number)
1) neg. control hTR	-	-
2) Standard 1 (8.6×10^6 copies)	8,518,000	18.98
3) Standard 2 (5.8×10^5 copies)	610,900	22.74
4) Standard 3 (6.9×10^4 copies)	61,260	26.02
5) Standard 4 (6.1×10^3 copies)	6,897	29.14
6) Standard 5 (4.4×10^2 copies)	420.1	33.13
7) pos. control hTR	994,400	22.05
8) Sample hTR	237,000	24.09
9) neg. control PBGD	-	-
10) pos. control PBGD	146,200	24.78
11) Sample PBGD	85,550	25.55

$$rel.\,hTR\ expression = \frac{Sample\ hTR}{Sample\ PBGD} = \frac{237,000}{85,550} = 2.77$$

Fig. 2. LightCycler run for hTR quantification. **a** Fluorescence using channel F2 / F1 is plotted against PCR cycle number. hTR and PBGD expression levels were determined for the commercial kit positive control (capillaries 7 and 10) and a sample of interest (capillaries 8 and 11). Five samples with known hTR copy-numbers were provided in the Kit and included as standards (capillaries 2–6). **b** Linear regression of the 5 standards was determined by using the 'Second Derivative Maximum' method. The linear regression and calculation of the correlation coefficient, r, serves to confirm accuracy and reproducibility. **c** The relative hTR expression in the Sample was determined as a ratio of hTR and PBGD levels which had been calculated from the standard curve

The cDNA is synthesized from the telomerase positive reference sample and diluted as follows: 1:1, 1:2, 1:4, 1:8, 1:16, 1:32, and 1:64. Aliquots of 10 µl are stored at −20°C.

A quantity of 2 µl of each dilution is needed for each LightCycler run as quantity standards. Therefore, to have sufficient material for experiments, 4 µl of cDNA should be produced for each expected LightCycler run.

Oligonucleotide Design

Pairs of primers are designed for hTERT, hTR and glyceraldehyde-3-phosphate dehydrogenase (GAPDH), used as a housekeeping gene (Table 1).

LightCycler PCR

GAPDH master mix for each 20-µl reaction:

	Volume [µl]	[Final]
LightCycler-DNA Master SYBR Green I	2	1×
$MgCl_2$ (25 mM)	1.6	3 mM
Primers (10 µM)	1+1	0.5 µM
H_2O (PCR grade)	12.4	
Total volume	18	

hTERT Master Mix for each 20-µl reaction:

	Volume [µl]	[Final]
LightCycler-DNA Master SYBR Green I	2	1×
$MgCl_2$ (25 mM)	0.8	2 mM
Primers (10 µM)	1+1	0.5 µM
TaqStart Antibody	0.16	
H_2O (PCR grade)	13.04	
Total volume	18	

hTR master mix for each 20-µl reaction:

	Volume [µl]	[Final]
LightCycler-DNA Master SYBR Green I	2	1×
$MgCl_2$ (25 mM)	1.6	3 mM
Primers (10 µM)	1+1	0.5 µM
TaqStart antibody	0.16	
H_2O (PCR grade)	12.24	
Total volume	18	

The relative GAPDH expression is determined prior to analyzing hTERT and hTR expression.

The procedure for semiquantitative PCR on the LightCycler is similar for each gene. A total of 18 µl of the respective master mix and 2 µl cDNA is added to each

Table 1. Oligonucleotides

hTERT (GenBank Accession #AF018167)				
	Position	Length	GC (%)	T_m (°C)
5'-CGGAAGAGTGTCTGGAGCAA-3'	1785	20	55.0	60.8
5'-CATGGACTACGTCGTGGGAG-3'	1980 R	20	60.0	61.0
Product	1785–1980	196		
hTR (GenBank Accession #AF047386)				
5'-CCTAACTGAGAAGGGCGTAGGC-3'	849	22	59.1	66.5
5'-CTAGAATGAACGGTGGAAGGCG-3'	961 R	22	54.5	65.3
Product	849–961	113		
GAPDH (GenBank Accession #M33197)				
5'-CACCCATGGCAAATTCCATGGC-3'	213	22	54.5	67.0
5'-GCATTGCTGATGATCTTGAGGCT-3'	509 R	23	47.8	65.4
Product	213–509	297		

glass capillary and placed in cooled adapters. The first eight capillaries are needed for one negative (H_2O-) control and the seven dilutions of the reference sample. Set the dilutions to "standard": 1:1=100; 1:2=50; 1:4=25; 1:8=12.5; 1:16=6.25; 1:32=3.125; 1:64=1.5625. The remaining capillaries can be used for the samples of interest. The sealed capillaries are briefly centrifuged (10 s at 700 g) and placed into the LightCycler rotor.

The following PCR protocol is used, with conditions for GAPDH/hTERT/hTR amplification as follows:
- Denaturation at 95°C for 2 min
- Amplification

Parameter	Value		
Cycles	40 (GAPDH)/45 (hTERT)/40 (hTR)		
Type	Quantification		
	Segment 1	Segment 2	Segment 3
Target temperature [°C]	95	63 (GAPDH)/ 60 (hTERT)/62 (hTR)	72
Incubation time [s]	1	5	14
Temperature transition rate [°C/s]	20	20	20
Acquisition mode	None	Single	None
Gains	F1=5; F2=10; F3=10		

- Melting Curve Analysis

Parameter	Value		
Cycles	1		
Type	Melting curves		
	Segment 1	Segment 2	Segment 3
Target temperature [°C]	95	70	95
Incubation time [s]	1	5	0
Temperature transition rate [°C/s]	20	20	0.1
Acquisition mode	None	None	Cont.
Gains	F1=5; F2=10; F3=10		

- Cooling at 40°C for 2 min

Results

The relative amounts of transcripts are calculated using the "second derivative maximum" method.

For this approach, the identity and specificity of the PCR product is confirmed by melting curve analysis which is part of the LightCycler analysis program. The specific melting point of the PCR product was correlated with its molecular weight as determined by agarose gel electrophoresis and fragment length analysis on an automated laser-fluorescence sequencer (ALFexpress, Pharmacia, Freiburg, Germany).

Error and Correlation

To obtain a high validity in the results, we accepted only a correlation coefficient (r) greater than or equal to -0.99. We recommend comparing only experiments using the same dilution series as standards.

The relative expression level can be calculated by dividing the amount of hTERT or hTR transcripts, respectively, by the amount of GAPDH transcripts [10, 11].

Comments

For samples where only a very small amount of RNA (e.g., from very small fine-needle biopsies and drug-treated cell cultures) could be obtained, we continued without DNase I digestion. To evaluate the effects of the remaining DNA contamination, we tested samples before and after purification. We observed no differences in hTERT expression with the hTERT kit.

Because the hTR sequence is also amplified from genomic DNA, we tested samples for hTR expression both with and without reverse transcriptase. For samples with high hTR expression, the amount of DNA product was very low (in the range of 5%) compared to the amount of mRNA product; therefore, we considered these to be suitable results. In cases with low hTR expression, where the quantities of

DNA and mRNA products were about the same, we chose to exclude these samples from further analysis.

In summary, we found that both the commercially available kits from Roche Diagnostics and our own protocol give reproducible and accurate results. However, for better reproducibility and comparability between results obtained in different laboratories, we recommend using the standardized kits.

References

1. Shay JW, Bacchetti S (1997) A survey of telomerase activity in human cancer. Eur J Cancer 33:787–791
2. Healy KC (1995) Telomere dynamics and telomerase activation in tumor progression: prospects for prognosis and therapy. Oncol Res 7:121–130
3. Poremba C, Bocker W, Willenbring H, Schafer KL, Otterbach F, Burger H et al. (1998) Telomerase activity in human proliferative breast lesions. Int J Oncol 12:641–648
4. Poremba C, Shroyer KR, Frost M, Diallo R, Fogt F, Schäfer K et al. (1999) Telomerase is a highly sensitive and specific molecular marker in fine-needle aspirates of breast lesions. J Clin Oncol 17:2020–2026
5. Poremba C, Willenbring H, Hero B, Christiansen H, Schafer KL, Brinkschmidt C et al. (1999) Telomerase activity distinguishes between neuroblastomas with good and poor prognosis. Ann Oncol 10:715–721
6. Morin GB (1989) The human telomere terminal transferase enzyme is a ribonucleoprotein that synthesizes TTAGGG repeats. Cell 59:521–529
7. Ramakrishnan S, Eppenberger U, Mueller H, Shinkai Y, Narayanan R (1998) Expression profile of the putative catalytic subunit of the telomerase gene. Cancer Res 58:622–625
8. Avilion AA, Piatyszek MA, Gupta J, Shay JW, Bacchetti S, Greider CW (1996) Human telomerase RNA and telomerase activity in immortal cell lines and tumor tissues. Cancer Res 56:645–650
9. Ito H, Kyo S, Kanaya T, Takakura M, Inoue M, Namiki M (1998) Expression of human telomerase subunits and correlation with telomerase activity in urothelial cancer. Clin Cancer Res 4:1603–1608
10. Poremba C, Scheel C, Hero B, Christiansen H, Schaefer KL, Nakayama J et al. (2000) Telomerase activity and telomerase subunits gene expression patterns in neuroblastoma: a molecular and immunohistochemical study establishing prognostic tools for fresh-frozen and paraffin-embedded tissues. J Clin Oncol 18:2582–2592
11. Poremba C, Hero B, Heine B, Scheel C, Schaefer KL, Christiansen H et al. (2000) Telomerase is a strong indicator for assessing the proneness to progression in neuroblastomas. Med Pediatr Oncol 35:651–655

Measurement of *MDR1* Gene Expression by Real-Time Quantitative RT-PCR Using the LightCycler Instrument

CHUNG-CHE CHANG*, SHERRIE PERKINS, CARL WITTWER

Introduction

Multidrug resistance protein (P-glycoprotein) is a 170-kDa membrane glycoprotein encoded by the *MDR1* gene. P-glycoprotein (P-gp) is believed to function as an ATP-dependent efflux pump for various toxins [1, 2].

P-gp has been found to be expressed at significant levels in the following normal tissues: the biliary canaliculi of liver, the proximal tubules of kidneys, small intestine, colon, and adrenal cortex. The expression of P-gp in these tissues has been believed to be one of the natural defense mechanisms that prevent cell damage from various toxins that these tissues may encounter. P-gp is also expressed in endothelial cells of the CNS, testes, and placenta and this type of expression is thought to contribute to blood–brain, blood–testicular, and blood–placental barriers. Some subsets of hematolymphoid cells also express P-gp, including bone marrow stem cells, lymphocytes, NK cells, and activated macrophages [1].

Overexpression of *MDR1* in cancer cells can cause cross-resistance to structurally unrelated categories of chemotherapeutic agents, known as multidrug resistance. *MDR1* expression in malignancies has the following patterns [1, 2]. First, some types of tumors commonly express high levels of *MDR1* at diagnosis. These cancers usually arise from tissues that normally express *MDR1*, such as colon, kidney, adrenal, pancreas, and liver cancer. Second, some types of tumors sometimes express high levels of *MDR1* at diagnosis but this is not common. These tumors include leukemias, lymphomas, chronic myelogenous leukemia in blast crisis, pediatric sarcomas, and neuroblastoma. Third, some types of tumors usually express low levels of *MDR1* or negative expression of *MDR1* at diagnosis. This group includes many chemotherapy-sensitive tumors such as untreated breast and ovarian cancer and small-cell lung cancers. Finally, in some tumors, including some breast and ovarian cancers, leukemias, lymphomas, and neuroblastoma, elevated *MDR1* expression is seen or acquired at relapse or recurrence. This has been thought to be due to either a selective process by chemotherapy or tumor progression independent of chemotherapy. *MDR1* gene promoter is a target for the c-Ha-RAS oncogene and the P53 tumor suppressor gene. Mutant P53

---

* Chung-Che Chang (✉) (e-mail: jeffchang@pol.net)
  Department of Pathology, Medical College of Wisconsin, Milwaukee, WI 53226, USA

can stimulate the *MDR1* promoter. C-*myc*, c-*fos* and c-*jun* appear to have a direct effect on *MDR1* expression [2].

Expression of *MDR1* has been shown to be an important prognostic indicator in adult acute myeloid leukemias (AML), non-Hodgkins' lymphoma, multiple myeloma, pediatric sarcomas involving soft tissue and bone, neuroblastoma, and small-cell lung cancer [3, 4]. Expression of *MDR1* in adult AML has been linked to lower remission induction rates and decreased remission duration [3].

A method of accurately and efficiently determining *MDR1* gene expression in malignancies is urgently needed because of the suggested prognostic importance of this marker [4, 5]. Additionally, clinical trials with P-gp modulators, such as PSC833 will also benefit from an improved assay of the MDR1 phenotype. We describe here a strategy to develop a real-time quantitative RT-PCR assay to measure *MDR1* mRNA. This assay will provide absolute value of the copy number of *MDR1* mRNA. Additionally, this assay provides a short turnaround time, about 40 min after RNA preparation, and is suitable for routine clinical application.

Materials

Equipment LightCycler instrument (Roche Molecular Biochemicals, Indianapolis, IN)

Reagents Amplification Primers (Idaho Technology, Salt Lake City, UT)
Hybridization Probes (Idaho Technology)
$MgCl_2$ Stock Solution (Idaho Technology)
dNTPs (2 mM stock containing dATP, dCTP, dGTP, dTTP)
Taq DNA polymerase (Promega, Madison, WI)
TaqStart antibody (ClonTech, Palo Alto, CA)
Enzyme diluent (10 mM Tris, pH 8.3, 250 µg/ml bovine serum albumin)

Procedure

Sample Preparation The total cellular RNA of the chemotherapy agent-sensitive cell line (MES-SA) and the chemotherapy agent-resistant cell line (MES-SA/DXA) was prepared using manual methods (RNAzol). Then cDNA was synthesized from 1 µg of total cellular RNA and 100 ng of random hexadeoxynucleotide primer (Pharmacia) in 10 µl of a solution containing 50 mM Tris, pH 8.3, 75 mM KCl, 3 mM $MgCl_2$, 10 mM dithiothreitol, 500 µM of each dNTP, and 10 units reverse transcriptase.

Primer Design The forward (5′ CAGGAGATAGGCTGGTTTGAXGT 3′, X is the modified T with Cy5 attached) and the reverse (5′ TTAGCTTCCAACCACGTGTAAATC 3′) primers were modified from published data to produce a 172-bp PCR product [6]. The hybridization probe (3′-fluorescein-labeled, 5′ GTCGGGTGTTAAGCTCCCCAA-CATCG 3′) were designed according to published guidelines [7, 8]. Briefly, the

Cy5-labeled primer and the fluorescein-labeled probe were designed to allow fluorescence resonance energy transfer (FRET) to occur when the hybridization probe annealed to the PCR products. The FRET was monitored by LightCycler in a real-time fashion (see "Results" for detail) for quantification purposes (Table 1).

Table 1. Oligonucleotides

MDR1 GenBank Accession #M29428				
	Region	Length	GC (%)	T_m (°C)
Primers				
CAGGAGATAGGCTGGTTTGAXGT[a]	151–173 (Exon 6)	23	48	64.6
TTAGCTTCCAACCACGTGTAAATC	840–863 (Exon 7)	24	42	63.8
Hybridization probe				
GTCGGGTGTTAAGCTCCCCAACATCG (3'-fluorescein-labeled)	177–202	26	58	71.6
Product		172		

[a] X, modified T with Cy5 attached.

LightCycler PCR

Hybridization Probe Master Mix:

	Volume [µl]	[Final]
dNTPs (2 mM stock)	1.0	200 µM
MgCl$_2$ (40 mM)	1.0	4 mM
Forward primer (5 µM each)	1.0	0.5 µM
Reverse primer (2 µM)	1.0	0.2 µM
Hybridization probe (1 µM)	1.0	0.1 µM
Diluted enzyme + TaqStart anibody[a]	1.0	0.4 U Taq and 8.8 ng TaqStart antibody
Sterile Glass Distilled H$_2$0	3.0	
Total volume	9	

[a] Diluted enzyme was prepared by adding 1 µl of Promega Taq (5 U/µl) to 1 µl of ClonTech TaqStart antibody (110 ng/µl). This mixture was incubated at room temperature for 5 min prior to the addition of 10.5 µl enzyme diluent. This solution was then added to the master mix.

In total, 9 µl of master mix and 1 µl of samples were added to the capillary tubes which were sealed, centrifuged in a microcentrifuge, and placed in the LightCycler rotor.

The following PCR protocol was used:
- Amplification

Parameter	Value		
Cycles	45		
Type	Quantification		
	Segment 1	Segment 2	Segment 3
Target temperature [°C]	94	50	72
Incubation time [s]	0	10	0
Temperature transition rate [°C/s]	20	20	1
Acquisition mode	None	Single	None
Gains	F1=2	F2=25	

- Melting Curve Analysis

Parameter	Value				
Cycles	1				
Type	Melting curve				
	Segment 1	Segment 2	Segment 3	Segment 4	Segment 5
Target temperature [°C]	94	74	67	60	84
Incubation time [s]	0	20	60	40	0
Temperature transition rate [°C/s]	20	20	20	20	0.2
Acquisition mode	None	None	None	None	Step

Establishing the Quantification Standard Curves for *MDR1* Using Different Known Amounts of Target PCR Products

The target PCR product for *MDR1* was obtained using cDNA from the resistant cell line (MES-SA/DXA) with known overexpression of *MDR1*. PCR was performed using the master mix listed above without the hybridization probe and using non-labeled forward primer. The product was amplified using the PCR cycle parameters as defined above. This final product was purified by CHROMA SPIN+TE-100 columns and quantified by spectrophotometry.

The standard curve for quantification of *MDR1* was established by using different known amounts of purified target PCR product of *MDR1* as templates for amplification. The samples contained 10^6, 10^5, 10^4, and 10^3 copies of purified target PCR products and were amplified with the master mix and PCR parameter as listed above.

Quantification of MDR1 Expression in Different Cell Lines

The cDNA was prepared with 1,700 cells from two cell lines: the chemotherapy agent-sensitive cell line (MES-SA) and the chemotherapy agent-resistant cell line (MES-SA/DXA). The cDNA was then amplified using the master matrix and PCR cycle parameters as listed above.

An appropriate negative control was included to confirm the specificity of PCR products. Caution was taken to avoid contamination from RNAase and PCR amplicons. For the prepared cDNA, β2 microglobulin, a housekeeping enzyme was amplified separately to examine the quality of the RNA retrieved.

Results

Quantification Standard Curves for *MDR1*

When the 3′-flourescein-labeled probe was annealed to the extension product of the Cy5-labeled primer, the fluorophores were brought into close enough contact for FRET to occur, resulting in increased fluorescence of Cy5 that reflects the cumulative amount of the products with advancing cycles. The PCR cycle number at which the fluorescence intensity of Cy5 became greater than the base line was defined as the "threshold cycle number" (Fig. 1). The quantification curve was established by plotting the log of the copy number of templates used for amplification versus the threshold cycle number. The results showed a linear inverse correlation (Fig. 2, $R^2=0.998$, $P<0.0015$).

During fluorescence melting curve analysis after the completion of the amplification, the probe then "melts off" at a characteristic temperature (melting peak or T_m). The T_m depends on the stability of the probe-product hybridized double strand structure. For a homogenous PCR product, the T_m can be used to assure the PCR products are the target products but not other contaminants. The specificity of the MDR1 products was verified by the T_m (68.1°C) immediately after the completion of the PCR reaction.

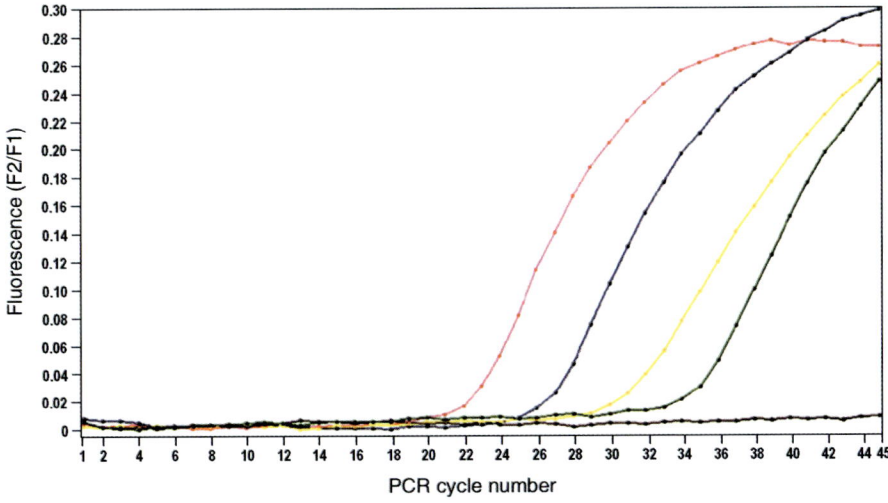

Fig. 1. The standard curve for quantification was established by using different known amounts of purified target PCR product of *MDR1* as templates for amplification. A The *red, blue, yellow,* and *green curves* contained 10^6, 10^5, 10^4, and 10^3 copies of purified target PCR product before amplification, respectively. The PCR cycle number at which the fluorescence intensity of Cy5 becomes greater than the baseline was defined as the "threshold cycle number". The *black curve* represents the water control

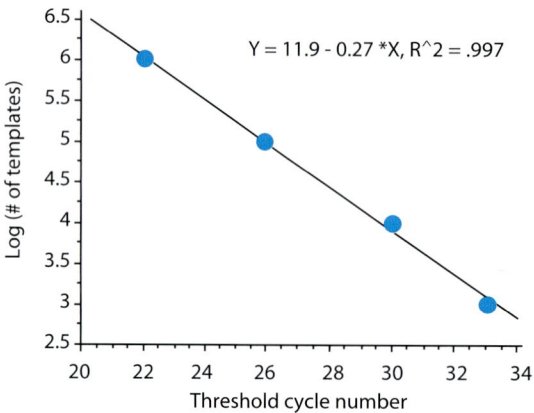

Fig. 2. The quantification curve was established by plotting the log of the copy number of templates used for amplification versus the threshold cycle number. The results showed a linear inverse correlation ($R^2=0.998$, $P<0.0015$)

Quantification of *MDR1* Expression in Different Cell Lines

The PCR reactions were performed as above using cDNA from the chemotherapeutic agent-resistant cell line (MES-SA/DX5) and the sensitive cell line (MES-SA). The threshold cycle number of samples was then determined from the PCR reaction. The standard curve established above was then used to calculate the copy number of *MDR1* mRNA molecules in each sample by extrapolation. The results indicated that a single cell from the resistant cell line contained 13 copies of *MDR1* mRNA and a single cell from the sensitive cell line was determined to contain one copy of *MDR1* mRNA. Therefore, there is a 13-fold difference in the expression of the *MDR1* gene between the chemotherapy agent-resistant cell line and the sensitive cell line.

Comment

Primer and Probe Design

The primers should have the T_ms as close to each other as possible for the best amplification efficacy. The hybridization probe should have a T_m about 5–10°C above that of primers to favor probe binding over primer binding.

Optimization of Assay

The most important factors for optimizing the assay are Mg^{++} concentration and the concentration of each primer and hybridization probe. Different primer ratios and different probe concentrations change the signal intensity generated from the PCR products (data not shown). Multistep annealing during the melting curve analysis enhances the signal intensity compared with one-step annealing (data not shown). Monitoring the ratio of the acceptor channel (F2) fluorescence intensity to fluorescein (F1) provides a smoother quantification curve than monitoring the F2 alone.

Conclusion

Our results indicate that this assay is a simple, reliable, and sensitive method for quantifying *MDR1* expression at the mRNA level with excellent turnaround time (about 45 min) after cDNA preparation. This method should be suitable for routine clinical use and for large series of oncological studies.

References

1. Ling V (1997) Multidrug resistance: molecular mechanisms and clinical relevance. Cancer Chemother Pharmacol 40[Suppl]:S3–S8
2. Chin KV et al. (1993) Function and regulation of the human multidrug resistance gene. Adv Cancer Res 60:157–180
3. Leith CP, Kopecky KJ, Godwin J et al. (1997) Acute myeloid leukemia in the elderly: assessment of multidrug resistance (MDR1) and cytogenetics distinguishes biologic subgroups with remarkably distinct responses to standard chemotherapy. A Southwest Oncology Group study. Blood 89:3323–3329
4. Chevillard S, Vielh PH, Validire P et al. (1997) French multicentric evaluation of MDR1 gene expression by RT-PCR in leukemia and solid tumors. Standardization of RT-PCR and preliminary comparisons between RT-PCR and immunohistochemistry in solid tumor. Leukemia 11:1095–1106
5. Beck WT, Grogan TM, Willman, CL et al. (1996) Method to detect P-glycoprotein-associated multidrug resistance in patients' tumors: Consensus recommendation. Cancer Res 56:3010–3020
6. Norgaard JM, Bukh A, Langkjer ST, Clausen N, Palshof T, Hokland P (1998) MDR1 gene expression and drug resistance of AML cells. Br J Haematol 100:534–540
7. Wittwer CT, Herrmann MG, Moss AA, Rasmussen RP (1997) Continuous fluorescence monitoring of rapid cycle DNA amplification. Biotechniques 20:130–138
8. Caplin BE, Rasmussen RP, Bernard PS, Wittwer CT (1999) LightCycler hybridization probes. Biochemica 1:5–8

Printing and Binding: Stürtz AG, Würzburg